INTRODUCTORY IMMUNOLOGY

INTRODUCTORY IMMUNOLOGY

Basic Concepts for Interdisciplinary Applications

SECOND EDITION

JEFFREY K. ACTOR

ELSEVIER

ACADEMIC PRESS
An imprint of Elsevier

Academic Press is an imprint of Elsevier
125 London Wall, London EC2Y 5AS, United Kingdom
525 B Street, Suite 1650, San Diego, CA 92101, United States
50 Hampshire Street, 5th Floor, Cambridge, MA 02139, United States
The Boulevard, Langford Lane, Kidlington, Oxford OX5 1GB, United Kingdom

Notices
Knowledge and best practice in this field are constantly changing. As new research
and experience broaden our understanding, changes in research methods, professional
practices, or medical treatment may become necessary.

Practitioners and researchers must always rely on their own experience and knowledge
in evaluating and using any information, methods, compounds, or experiments described
herein. In using such information or methods they should be mindful of their own safety and
the safety of others, including parties for whom they have a professional responsibility.

To the fullest extent of the law, neither the Publisher nor the authors, contributors, or
editors, assume any liability for any injury and/or damage to persons or property as a
matter of products liability, negligence or otherwise, or from any use or operation of any
methods, products, instructions, or ideas contained in the material herein.

Library of Congress Cataloging-in-Publication Data
A catalog record for this book is available from the Library of Congress

British Library Cataloguing-in-Publication Data
A catalogue record for this book is available from the British Library

ISBN : 978-0-12-816572-0

For information on all Academic Press publications
visit our website at https://www.elsevier.com/books-and-journals

Working together
to grow libraries in
developing countries

www.elsevier.com • www.bookaid.org

Publisher: Andre G. Wolff
Acquisition Editor: Linda Versteeg-Buschman
Editorial Project Manager: Timothy Bennett
Production Project Manager: Maria Bernard
Designer: Christian Bilbow

Typeset by SPi Global, India

Dedication

To my father, Paul Actor, PhD, who instilled in me a sense of excitement about the wonders of science and the curiosity to seek questions about how biological systems function; and to my mother, Ruthe Actor, who taught me to seek value in everything I accomplish and to approach all challenges with an open mind.

Contents

7. Autoimmunity: Regulation of Response to Self

8. Immune Hypersensitivities

9. Vaccines and Immunotherapy

10. Cancer Immunology

Preface

Our bodies have evolved a protective set of mechanisms, comprised of cells and organs, as a primary defense to maintain health. In essence, we have developed internal tools to preserve health and homeostasis. Indeed, a working definition of *health* embraces the effective elimination or control of life-threatening agents. This includes both infectious agents attacking from the outside and internal threats, such as tumors. Immune responses, therefore, are designed to interact with, and respond to, the environment to protect the host against pathogenic invaders and internal dangers. The goal of this book is to appreciate the components of the human immune system that work together to confer protection.

We will begin our discussion by establishing a foundation for subsequent chapters, through the presentation of the systems and cells involved in immune responses. Chapter 1 will give a general overview on the mechanisms in place to fight against disease. Components and pathways will be defined to allow the presentation of concepts of innate (always present) and adaptive (inducible and specific) responses, and how these responses interact with one another to form the basis for everyday protection. These concepts will form the foundation to examine the process of defense against various classes of pathogens. Chapter 2 will examine the coordinated effort of cells and blood components in the development of inflammation as it is related to protection against infection. Chapter 3 will introduce the basis for the function of adaptive components, exploring the generation of B lymphocytes and the nature of antibodies. Chapter 4 will extend this discussion to T-lymphocyte populations and examine how they serve as ringleaders for immune function. Chapter 5 will discuss immune responses, with an element of detail focused on commonly encountered infectious organisms. This overview also will include how the initial engagement of pathogens by innate components triggers pathways that cause inflammation. A section of this chapter will introduce the concept of opportunistic infections and diminished response when individuals are immunocompromised.

Effective immune surveillance is paramount to maintaining health. Chapter 6 will examine the basic disorders of immune function. Too little of a response results in an inability to control threats, which thus is ineffective to eliminate infectious agents. This lack of reactivity (*hyporeactivity*) leads to holes in our immune repertoire. This may be the result of genetic deficiencies or acquired compromise of immune function. In the same manner, responses representing excessive activity also can lead to damage to the host. This overaggressive response, a state of *hyperreactivity*, may reflect a productive response that increases in intensity and duration without effective control. The dysregulation leads to tissue-damaging events and eventual states of disease.

The chief function of the immune system is to distinguish between what is you (self) and what constitutes external threats. When the ability to distinguish these elements is compromised, autoimmunity may arise. In Chapter 7, autoimmune dysfunction will be addressed, moving from basic concepts to the specific mechanisms involved in major clinical disorders. This includes a detailed discussion of how the self is recognized, as well as the mechanisms involved in tolerance that limit reactivity to our own tissues. The goals here are to present the clinical manifestation of autoimmunity in such a manner that outward symptoms are understood through investigation of the molecular targets involved in the host immune self-recognition response. At other times, misdirected recognition of nonself elements, such as environmental allergens that typically are considered harmless, result in the development of clinical presentations. Therefore, Chapter 8 will examine the processes involved in the manifestation of immune dysfunction, including the concepts of immune hypersensitivities that lead to clinical disease.

The general topic of vaccines will be addressed in Chapter 9, including both how they work and a frank discussion of the relative truths and myths surrounding their use. This chapter also will contain information on newly developed therapeutics that are grounded in methods that lead to immune modification and factors that promote a healthy immune response (for example, lifestyle activities and good common practices). Indeed, it is critical that we maintain a healthy balance throughout our lives to ensure functional immune response as we age. The challenges faced at each stage of our lives, from that found in the prenatal/newborn period, to midlife, to so-called maturity, are mentioned in a way that encourages a healthy condition to allow the optimization of immune function.

A discussion of natural (effective) response to tumor development in Chapter 10 will lead to an investigation into the components of immune function to eliminate potentially dangerous precancerous events naturally. This will be followed by coverage of the challenges that we face when protective responses fail and tumors develop. A section also will contain information on cancers of the immune system and the problems that arise when the protective cells themselves become the cause of tumorigenic activity.

Chapter 11 will delve once more into details underlying the concepts of "self" versus "nonself" and blood types, with the goal to present genetic relationships (similarities as well as differences) between individuals. The mechanisms of the immunobiology of transplantation will be discussed, with details about the contributing cells and factors involved in transplant acceptance and rejection. The challenge is to appreciate the importance of innate and adaptive components in graft recognition, as well as to recognize the clinical consequences of transplantation that affect aspects of daily activities. Rejection topics will be discussed, including graft-versus-host disease (GVHD), as well as modern immune-based therapeutics designed to alter immune function to limit graft rejection.

Finally, additional information and resources will be provided in Chapter 12 to allow readers to develop an immune-based foundation of knowledge to understand the clinical tests associated with identifying immune parameters that arise during development of disease states. As such, this includes an introduction to mechanisms that form the basis of immune-related diagnostics and identification of immune properties of the blood during disorders.

All in all, the hope is that this book will present the concept of the immune system so that readers may better understand immune-based diseases resulting from either immune system component deficiencies or excess activity. This book is aimed at those who want to know more, to encourage readers to explore deeper. It is aimed at the curious who have never previously considered the underlying facets of effective immune function. To students who wish to expand their basic knowledge of biological systems. To physicians seeking to refresh their understanding of the immune concepts that cause clinical disease. To nurses who desire to expand their view of symptom development in patients. To patients who want a simple explanation for the complex way that their bodies respond in the context of the world they live in. And finally, to all who seek to find out how the body confers protection against infectious agents, maintains everyday homeostasis, and guards against dysregulation of the normal response to confer health and control the development of disease.

Acknowledgments

I would like to give a special thank-you to Keri Smith, PhD, for academic contribution to the chapters on antibodies and immunoassays. In addition, thanks goes to Robert L. Hunter, MD, PhD, for sharing his positive outlook on life and showing me the joy of embracing scientific thought; to my children, Jonas and Amanda, for sharing a thirst for knowledge and for continuing to ask questions; and to my wife, Lori, for her love and patience and for understanding my desire to complete this project. Finally, I extend gratitude to all the students I've mentored at the McGovern Medical School, for their feedback and suggestions toward completion of this text.

1

A Functional Overview of the Immune System and Immune Components

CHAPTER FOCUS

To establish a foundation to appreciate how components of the immune system work together to protect against the development of clinical disease. The basic systems and cells involved in immune responses will be presented in this chapter to give a general overview of functional immunity. Components and systems will be defined to allow an understanding of concepts of innate (always present) and adaptive (inducible and specific) responses, and how these responses interact with one another to form the basis for protection against disease.

IMMUNE HOMEOSTASIS

A functional immune system offers constant surveillance of human beings in relationship to the world. It confers a balanced state of health through effective elimination of infectious agents (bacteria, viruses, fungi, and parasites) and through control of malignancies. Indeed, the immune system has evolved to allow cells and organs to interact with the environment to protect against harmful invaders. At the same time, mechanisms are in place to instill tolerance toward the naturally occurring microbiome (i.e., microbial and viral agents) that reside within us in symbiotic ways. Taken together, these responses represent a balance of components that ward off the development of clinical disease.

SELF VERSUS NONSELF

Discrimination between the self and the nonself is considered the chief function of the immune system. We are under constant assault by invaders. Our bodies represent prime substrates for organisms to grow and reside, with an abundance of nutrients, warmth, and protection from the outside elements. The immune system is basically a series of obstacles to limit and inhibit pathogen entry and then to attack and destroy those organisms once they enter the body. The immune response is exquisitely designed to recognize these invaders as "foreign." In fact, the major feature that renders our immune system so effective is its ability to distinguish our body's own cells (the **self**) from that which it considers foreign (termed the **nonself**). Each of our cells carries specific tags, or molecular markers, that label it as "self." These markers are important, as they not only determine what is unique about us, but they also distinguish one person from another.

Almost anything and everything that registers as "nonself" will trigger an immune response. An intricate system of molecular communication and cellular interactions allows immune components to function in concert to combat disease-causing organisms. The foreign agent (microbe, virus, parasite, etc.), or any part of it that can be specifically recognized, is called an *antigen*. Simply put, an **antigen** is defined as any substance or physical structure that can be recognized by the immune system. Major classes of antigens include proteins, carbohydrates, lipids, and nucleic acids. If an antigen is of high complexity and weight, it can trigger full immune activity and become **immunogenic**.

The ability to distinguish our own cells from the outside world is critical to maintaining functional protection. If this ability is lost (e.g., when "self" tissue is seen as foreign), then our immune system launches an aggressive response against our own tissues. This is what happens during **autoimmunity**, where destruction of the self leads to clinical disease.

The immune system maintains a balance of responsiveness. Too little a response is ineffective, while too aggressive a response can lead to targeted destruction of bystander tissues. Both scenarios are equally as devastating and may result in clinical disease. The regulation of immune function and overall immuno-homeostasis is under the control of multiple factors, including genetic components and environmental cues. The intensity and duration of the responses must be sufficient to protect against invading pathogens, with prompt and specific downregulation when the foreign material (the antigen or the pathogen) is no longer present. The clinical state that arises when immune responses are not properly regulated is termed **hypersensitivity**, a state of excessive or inappropriate responses leading to disease. As one might imagine, hypersensitivity can occur in many different forms, depending upon which arm of the immune system is dysregulated.

INNATE AND ADAPTIVE IMMUNITY

The immune system is loosely divided into two major functional categories termed **innate** and **adaptive immunity**. Innate immune mechanisms provide the first line of defense from infectious diseases (Table 1.1). The innate immune components exist from birth and consist of components available prior to the onset of infection. These defensive components include both physical barriers and biochemical factors. Defensive innate mechanisms may be anatomic (skin, mucous membranes), physiologic (temperature, low pH, chemical mediators), phagocytic (digestion of microorganisms), or inflammatory (vascular fluid leakage).

Innate mechanisms are particularly powerful at limiting infections. However, once the infectious agent is established inside the body, a more focused set of reactive molecules and cellular components are required to combat the specific organism. An intricate system of molecular communication and cellular contact allows the components of the innate immune

TABLE 1.1   Innate Defensive Components

Component	Effectors	Function
Anatomic and physiologic barriers	Skin and mucous membranes	• Physical barriers to limit the entry, spread, and replication of pathogens
	Temperature, acidic pH, lactic acid	
	Chemical mediators	
Inflammatory mediators	Complement	• Direct lysis of pathogen or infected cells
	Cytokines and interferons	• Activation of other immune components •
	Lysozymes, defensins	• Bacterial destruction
	Acute phase proteins and lactoferrin	• Mediation of response
	Leukotrienes and prostaglandins	• Vasodilation and increased vascular permeability
Cellular components	• Polymorphonuclear cells • Neutrophils, eosinophils • Basophils, mast cells	• Phagocytosis and intracellular destruction of microorganisms
	• Phagocytic-ndocytic cellsMonocytes and macrophages • DCs	• Presentation of foreign antigens to lymphocytes

group to trigger the cells involved in adaptive immunity. In essence, both innate and adaptive components must function in concert to combat and control disease successfully.

Adaptive (also called **acquired**) immune responses account for specificity in the recognition of foreign antigenic substances. It is critical to understand that specificity of the adaptive immune response lies within two distinct subsets of white blood cells, called **lymphocytes**. Lymphocyte recognition of unique shapes associated with foreign antigens is accomplished by functional receptors residing on their cellular surface. Key elements of the acquired immune responses are compared to functional elements of the innate response, as listed in Table 1.2.

The adaptive immune response is subdivided into functional groups representing **humoral** and **cellular immunity**, based on the participation of two major cell types. Humoral immunity involves **B lymphocytes** (also called **B cells**), which synthesize and secrete **antibodies**. Cellular immunity (also called **cell-mediated immunity**) involves effector **T lymphocytes** (also called **T cells**), which secrete immune regulatory factors following interaction with specialized processing cells (called **antigen presenting cells,** or **APC**s) that show the lymphocytes foreign material in the context of self-molecules.

TABLE 1.2   Key Elements of Innate and Acquired Immune Responses

Innate	Adaptive
Rapid response (minutes to hours)	Slow response (days to weeks)
Polymorphonuclear leukocytes (PMNs) and phagocytes	B cells and T cells
NK cells	NKT cells
Preformed effectors with limited variability	B-cell and T-cell receptors with highly selective specificities to foreign agents
Pattern recognition molecules recognizing structural motifs	
Soluble activators	Antibodies (humoral)
Proinflammatory mediators	Cytokines (cellular)
Nonspecific	Specific
No memory, no increase in response to secondary exposure	Memory, maturation of secondary response to reexposure

ANATOMY OF THE IMMUNE SYSTEM

The immune system is just that: a system. It is a network of protective barriers, organs, cells, and molecules. Specifically, there are subsets of primarily bone-marrow-derived cells that circulate throughout the body. Indeed, the power of the system is that contributing immune cells can be found within every major organ and every tissue. These cells are available to be called into action at very short notice. The interactions are managed by a series of central **lymphoid organs** (e.g., bone marrow, thymus, spleen, and lymph nodes) containing high levels of lymphocytes. The immune-based lymphoid organs are where leukocytes of myeloid and lymphoid origin mature, differentiate, and multiply (Fig. 1.1). Cells also

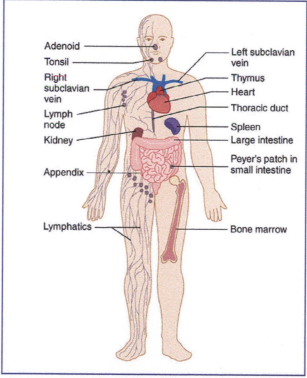

FIG. 1.1 Distribution of lymphoid tissues. Primary lymphoid organs, such as the bone marrow and thymus, are the major sites of lymphopoiesis (lymphoid hematopoiesis) and the locales where lymphocytes differentiate. Secondary lymphoid organs, such as the spleen and lymph nodes, are sites where antigen-driven proliferation and maturation of lymphocytes occur.

accumulate outside these major organs, residing in less defined areas (e.g., throughout the gut or skin), to allow protective responses at local sites when rapid responses are needed.

Primary lymphoid organs are the sites where lymphoid cells are generated. This act of cellular development, or **lymphopoiesis**, occurs in the liver in the fetus, and then in the bone marrow after birth. Islands of progenitor stem cells give rise to immune system cells that are subsequently released into the blood. A specialized primary lymphoid organ is the **thymus**, a pyramid-shaped gland that is located beneath the breastbone at the same level as the heart (Fig. 1.2). Immature lymphocytes leave the bone marrow and circulate directly through the thymus. The principal function of the thymus gland is to educate T lymphocytes to distinguish between what is you (self) and what is not (nonself). When lymphocytes leave the thymus, they are committed along a certain pathway of activity and are ready to perform their effector functions. The other set of lymphoid organs, called **secondary lymphoid** organs, are structured compartments that provide a favorable environment for cell contact and activation of committed cells.

Lymphocytes continuously leave the blood vessels, migrating throughout the body, in which they perform surveillance activities. The return path to circulation begins when cells mix with fluids that naturally bathe tissues. The mixture eventually drains through small vessels to return materials to the blood supply. The fluid in the tissue is called **lymph**, and it carries cells and debris through small-vessel **lymphatics**, which direct the lymph and cells through secondary lymphoid organs before reaching the thoracic duct, where fluid and cells are returned to the venous circulation of the blood supply.

Lymph nodes are focal nodules connected via the draining lymphatic highway. They are placed throughout the body, with groupings found in the groin, armpits, and abdomen. They represent local nodes for antigen and cellular drainage. It is here where lymphocytes can interact and communicate with APCs, allowing local presentation of antigenic particulates found in nearby regions of the body. Think of the regional lymph nodes as rest stops along the highway, where cells can mingle and discuss local and systemwide information. If there is a need for immediate response, cells can actively mobilize efforts to defend or repair tissues. In essence, this is where antigen-driven proliferation and differentiation occur. Local lymph nodes become swollen and painful as cells respond to regional damage and drained materials, consequences of the activation of the immune response team. Within the lymph nodes, the areas of response are called **germinal centers**. Indeed, this term is used to describe any local foci of responding lymphocytes in secondary immune reactive sites.

Just as the lymph nodes are connecting nodes for the lymphatics, the **spleen** is a filtering organ for circulating blood. The spleen, located in

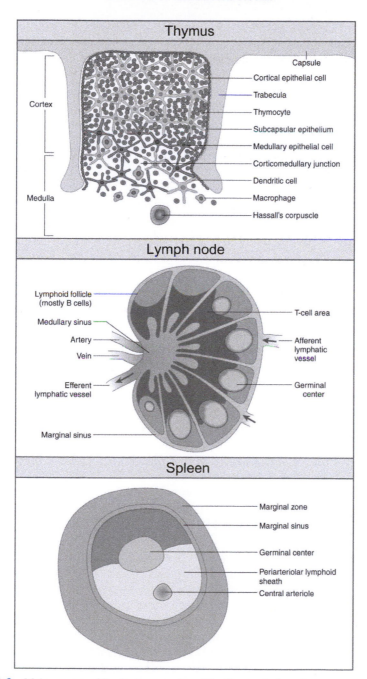

FIG. 1.2 Major organs of the immune system. The thymus is the primary organ responsible for the education of lymphocytes to differentiate between the self and the nonself. The lymph nodes are secondary organs placed throughout the body, functioning as focal nodules where lymphocytes interact and communicate with APCs. The spleen is a secondary organ, where resting lymphocytes reside to readily mobilize in response to the detection of foreign materials.

the upper portion of the abdomen, can be considered a holding facility where both innate and resting adaptive cells reside in specialized compartments. The main areas of the tissue are comprised of either lymphoid cells (called the "white pulp"), where immune cells interact, or red blood cells (RBCs) and associated areas where RBCs flow (called the "red pulp"). These compartments for cellular activation allow cells to readily activate and mobilize in response to communication signals produced as an indication that foreign materials have been identified.

Another major secondary immune organ is not a truly defined organ; rather, it is comprised of tissue areas of loosely associated cellular aggregations where contact with foreign material is common. This type of immune aggregation is found in tissue layers that line the intestines, the lung, and the nasal cavities. These aggregates are called **mucosa-associated lymphoid tissue (MALT)**, and represent areas of rapid surveillance and detection for organisms entering through major openings in our bodies. The tonsils, adenoids, appendix, and Peyer's patches (organized tissue in the large intestines) represent a more formal association of parenchyma that shares the same functional parameters as MALT. In an analogous manner, aggregates lining the bronchial regions are called **bronchial/tracheal-associated lymphoid tissue (BALT)**; in cases of chronic/persistent pulmonary infection, these structures can develop further into **inducible BALT (iBALT)**, where distinct regions of follicles and germinal centers containing specialized immune cells can be identified. Finally, aggregates that line the intestinal tract are referred to as **gut-associated lymphoid tissue (GALT)**. There are specialized cells in some of these aggregates; **M cells**, or *microfold cells*, can be found in the follicle-associated epithelium of Peyer's patches. M cells sample antigen from the lumen of the small intestine and deliver it via transcytosis to immune cells located on their basolateral side.

CELLS OF THE IMMUNE SYSTEM

Leukocytes is the term given to white blood cells that play a functional role in either innate or adaptive responses. This population of cells can be broken into two main groups, referred to as **myeloid** or **lymphoid** cells, depending on which developmental path was taken by the stem cells in the bone marrow during development (Fig. 1.3). Myeloid cells are considered as the first line of defense and thus constitute the major cell types involved in innate immunity (Table 1.3). Myeloid cells include highly phagocytic, motile **neutrophils, monocytes**, and **macrophages**, as well as **dendritic cells (DCs)**, which provide relatively immediate protection against most pathogens. The other myeloid cells, including **eosinophils, basophils**, and their tissue counterparts, **mast cells**, are involved in the

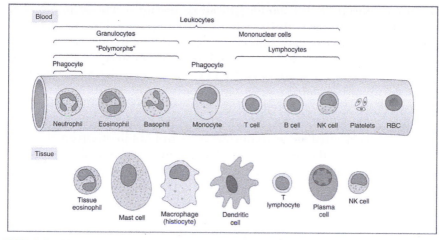

FIG. 1.3 Nomenclature and location of immune system cells.

TABLE 1.3 Myeloid Leukocytes and Their Properties

Phenotype	Morphology	Circulating differential cell count[a]	Effector function
Neutrophil	Polymorphonuclear leukocyte (PMN) granulocyte	$2–7.5 \times 10^9/L$	Phagocytosis and digestion of microbes
Eosinophil	PMN granulocyte	$0.04–0.44 \times 10^9/L$	Immediate hypersensitivity (allergic) reactions; defense against helminths
Basophil	PMN granulocyte	$0–0.1 \times 10^9/L$	Immediate hypersensitivity (allergic) reactions
Mast cell	PMN granulocyte	Tissue-specific	Immediate hypersensitivity (allergic) reactions
Monocytes	Monocytic	$0.2–0.8 \times 10^9/L$	Circulating macrophage precursor
Macrophage	Monocytic	Tissue-specific	Phagocytosis and digestion of microbes; antigen presentation to T cells
DC	Monocytic	Tissue-specific	Antigen presentation to naïve T cells; initiation of adaptive responses

[a] Normal range for 95% of population, ± standard deviations.

defense against parasites and the genesis of allergic reactions. In contrast, lymphoid cell types include those that mediate specific immunity; simply put, they are the cell types (**phenotypes**) that have defined receptors to physically interact with and recognize foreign materials (Table 1.4). These types of cells fall into the acquired category and include the B lymphocytes and the T lymphocytes that were mentioned previously.

In addition, a group of cells called **natural killer T (NKT) cells** exist, which is a specialized subset of lymphocytes. A functionally related set of lymphoid cells that are also considered lymphoid in origin are the **natural killer (NK) cells**. Although similar in name, and slightly confusing, NK cells are distinct from NKT cells.

FIRST-LINE DEFENDERS: THE MYELOID CELLS

Neutrophils are highly adherent, motile, phagocytic leukocytes that are typically the first cells recruited to acute inflammatory sites. Neutrophils are the most abundant of the myeloid populations. They are stored in the bone marrow and are readily released during infection. Neutrophils engulf (**phagocytose**) and devour pathogens, after which they use specialized destructive "granular" enzymes and toxic molecules to kill the ingested organism. Many of these enzymes regulate reactive oxygen species, such as superoxide and nitric oxide, to mediate killing. This **respiratory burst**, the phase of elevated oxygen consumption shortly after the cellular ingestion of organisms, allows the neutrophil to limit expansion and pathogen growth. Neutrophils also defend through the release of DNA-chromatin structures, described as **neutrophil extracellular traps (NETs)**. Released proteins adhere to the NETs, allowing a scaffold of enzymes to function in extracellular spaces to combat microorganisms. The process of creating NETs is called *netosis*. In essence, the neutrophil

TABLE 1.4   Lymphoid Leukocytes and Their Properties

Total lymphocytes 1.3–3.5×10⁹/L			Effector function
B cell	Monocytic	Adaptive	Humoral immunity
Plasma cell	Monocytic	Adaptive	Terminally differentiated, antibody-secreting B cell
T cell	Monocytic	Adaptive	Cell-mediated immunity, immune-response regulation
NKT cell	Monocytic	Adaptive	Cell-mediated immunity (glycolipids)
NK cell	Monocytic	Innate	Innate response to microbial or viral infection

acts as a first responder, giving immediate help and slowing the spread of infection while the next phases of the immune response are mobilized.

Like the neutrophil, **eosinophils** also have specialized molecules, stored as granules, which they release in defensive response to infection. While the neutrophil is adept at engulfing smaller organisms, the eosinophil is successful against large multicellular parasites that are too big to fit inside any one cell. They primarily defend against extracellular targets.

Basophils, and their tissue counterparts, **mast cells**, produce **cytokines,** which help defend against parasites. However, these cells are best known clinically for their role in allergic inflammation. Basophils and mast cells display surface membrane receptors for a specific class of antibodies; they release a host of molecules, such as histamine and vascular mediators that affect blood flow, when cell-bound antibodies recognize allergens.

Macrophages are also involved in phagocytosis and the intracellular killing of microorganisms. Macrophages can reside for long periods in tissues (outside the bloodstream). These cells are highly adherent, motile, and phagocytic; they marshal and regulate other cells of the immune system, such as T lymphocytes. Macrophages may be classified further based on the subsets of molecular mediators that they secrete. The "classical" M1 macrophage produces proinflammatory mediators to augment T lymphocytes to protect against bacteria and viruses. The M2 macrophages represent a broader phenotypic subset that elicit a different set of molecules; the M2 macrophages are more closely involved with larger parasite control, wound healing and tissue repair.

In a similar manner to the macrophage, DCs also provide a critical link between innate and adaptive immunity by interacting with T cells. DCs originate from cell precursors on an independent path from macrophages. The unique function of DCs lies in the manner of how they deliver strong signals for the development of memory responses. DCs recognize foreign agents through a series of unique receptors that recognize general shapes (motifs) on foreign organisms. Different cellular subsets exist (myeloid-conventional DCs and plasmacytoid DCs), enabling this group of cells to both prime and dictate how subsequent responses will develop. Both phenotypes of DCs are equipped with "molecular sensors" to recognize foreign pathogens, with the innate ability to understand environmental cues in the form of molecular mediators. This permits them to exquisitely guide the specificity and magnitude of subsequent immune responses.

Macrophages and DCs are aptly called **APCs**. After they destroy pathogens, they show (present) chopped pieces to T lymphocytes, thereby mediating a connection with the adaptive immune response. Basically, the pathogen is digested inside the presenting cell, and small fragments are shown on their cell surface for recognition by adaptive lymphocytes. Indeed, the macrophages and DCs, along with other specialized APCs, contain a myriad of surface molecules that directionally drive responsiveness and focus adaptive responses to allow productive elimination of microorganisms.

Finally, **platelets** and **erythrocytes** (RBCs) also arise from bone-marrow-derived myeloid megakaryocyte precursors. Platelets are involved in blood clotting (coagulation) and wound repair. During the process of wound repair, they release inflammatory mediators involved in innate immune activation. Of interest, recent findings show that in addition to their major role in oxygen exchange, RBCs can serve as dynamic reservoirs of inflammatory mediators.

ADAPTIVE AVENGERS: THE LYMPHOID CELLS

Lymphoid cells provide efficient, specific, and long-lasting immunity against microbes and pathogens and are responsible for acquired immunity. As a group, they respond to infectious invasion only after the myeloid cells have begun their work. Indeed, the innate responding cells discussed in this chapter send signals to the lymphoid population to "stop and smell the inflammation" and engage them to respond specifically, in a directional manner. Lymphocytes differentiate into separate lineages. The B lymphocytes secrete antibodies. The T lymphocytes operate in a supervising role to mediate cellular and humoral immunity. NK cells are critical in defense against viral agents. B and T lymphocytes produce and express specific receptors for antigens, while NK cells do not.

LYMPHOCYTES

The adaptive immune response accounts for specificity in recognition of foreign substances, or antigens, by functional receptors residing on their cell surfaces (Fig. 1.4). The **B-cell antigen receptor** is the surface **immunoglobulin**, an integral membrane protein with unique regions to bind specific antigenic shapes. There can be thousands of identical copies of the receptor present on the surface of a single B lymphocyte (simply called a **B cell**). B-cell activation occurs when the receptor encounters the antigen. This leads to a morphological change in B cells, which now multiply to become secretory factories to make and release soluble immunoglobulins, or **antibodies**. B cells that actively secrete antibodies are called **plasma cells**. We will see in subsequent chapters how antibodies are critical to target and neutralize invading organisms.

The T lymphocyte matures as it passes through the thymus. Like its B-cell counterpart, it also has a surface receptor, the T-cell receptor, called the **TCR**. The TCR is structurally similar to the antibody; it too recognizes

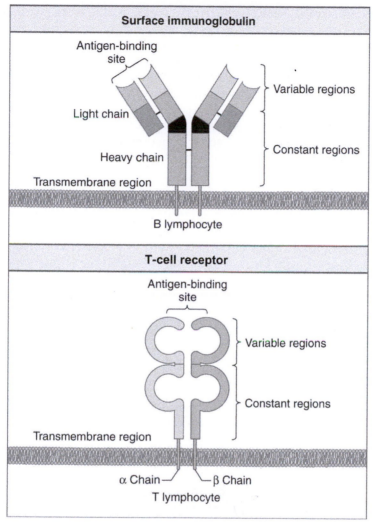

FIG. 1.4 Basic structure of the antigen receptors. Depicted here are the antigen receptor on the surface of a B cell (the immunoglobulin B-cell receptor; BCR) and on the T cell (the TCR).

specific pieces of the antigen, although we will see in later chapters that there are restrictive constraints on this interactive process. Unlike the antibody, the TCR is present only on its surface and is not secreted when T cells are activated. The process of T cell activation is quite complex and requires a group of cells to show pieces of the antigen physically to the T cell. These assisting cells are the APCs described here. The surface molecules on the presenting cell which show antigen to the T cells are also involved as tags for "self" identity. These molecules are called **major**

histocompatibility complex (MHC) molecules. Together, these cellular events form a regulated pathway to present foreign antigens for subsequent recognition and triggering of specific responses to protect against disease.

The T lymphocytes are the true ringleaders of the adaptive response. Different subsets control different functions. Some help B cells produce antibodies. Others help the myeloid cells become more efficient at destroying pathogens. Some preferentially function to kill viral infected cells or tumor cell targets. Finally, some function as regulators, conferring immune tolerance or establishing limits on responsiveness.

CLUSTER OF DIFFERENTIATION

The **cluster of differentiation (CD)** designation refers to proteins found on the surface of cells. Each unique surface molecule is assigned a different number, which allows cell phenotypes to be identified. Surface expression of a particular CD molecule is useful for the characterization of cell phenotypes. CD molecules may be specific for just one cell or cell lineage, or they may not. The CD designation will be used throughout this book. The official listing of determinants has identified more than 400 individual and unique markers. A simplified list of CD molecules and their associated cell function is given in Table 1.5. CD-specific diagnostic agents have been useful for determining the functions of proteins and for identifying their distribution in different cell populations. They also have been useful for measuring changes in the proportion of cells carrying those markers in patients with disease.

SUMMARY

- Our immune responses are designed to interact with the environment to protect against pathogenic (disease-causing) invaders.
- Immunity is based on functional discernment between self and nonself, a process that begins in utero and continues through adult life.
- A network of organs allows the generation of functional cells and regional centers for cellular interaction and recognition of foreign (nonself) material.
- Physical barriers comprise the first line of defense against pathogens. Once pathogens invade the body, innate systems react on short notice to confer protection. Cells of the myeloid lineage (including neutrophils and macrophages) are readily available to allow the containment of

TABLE 1.5 Selected CD Markers and Associated Functions

CD marker	Biological function
CD1	Presentation of glycolipids to NKT cells
CD2	T-cell adhesion molecule
CD3	Signaling chains associated with the TCR
CD4	Coreceptor for Class II MHC on T cells
CD8	Coreceptor for Class I MHC on T cells
CD11	Leukocyte adhesion
CD14	Lipopolysaccharides (LPS) binding protein receptor
CD18	Beta-2 integrin
CD19	B-cell signal transduction
CD20	B-cell calcium channel activation
CD21	B-cell activation
CD25	Interleukin-2 (IL-2) receptor alpha-chain
CD28	T-cell costimulatory molecule
CD32	Immunoglobulin G (IgG) receptor
CD34	Hematopoietic stem cell marker
CD40	Class switching on B cells
CD44	Lymphocyte adhesion
CD45	Lymphocyte activation; memory marker
CD54	Adhesion molecule
CD56	NK cell marker
CD58	Adhesion molecule
CD59	Regulator of complement membrane attack complex assembly
CD62L	T-cell adhesion to high endothelial venules
CD80	Costimulatory receptor on APCs
CD86	Costimulatory receptor on APCs
CD95	Induction of apoptosis
CD152	Negative regulator for T cells
CD154	Involved in B-cell proliferation and class switching

organisms until the specific adaptive immune response becomes engaged.

- Innate cells form a primary network to set the stage for the arrival of adaptive cells. The macrophages and DCs serve as a trigger to present foreign materials and give directionality to the next stage of adaptive cellular responses.
- Adaptive immunity allows discrimination of antigens using defined receptors. B lymphocytes make antibodies as a major defensive strategy. T lymphocytes perform as the functional ringleaders to direct the defensive response.

The Inflammatory Response

CHAPTER FOCUS

To examine the coordinated effort of cells and blood components to elicit the inflammatory response. Information will be presented to allow an understanding of how initial organism detection launches an inflammatory cascade. Factors that create redness and swelling, heat, and pain will be addressed, with the overall goal to identify and appreciate the processes involved in the development of inflammation as related to protection against infectious disease.

INFLAMMATION

Innate immune mechanisms provide the first line of defense against infectious disease. At the very basis of this response is a set of components that are continuously available for immediate response and are present prior to the onset of infection or tissue damage. These constitutive factors initiate a series of coordinated events that allow rapid detection of microorganisms and then trigger a signaling cascade to call in cell types that are adept at controlling infection. While these events productively begin both healing and protective processes, they also culminate in inflammation.

Greek literature suggests that ancient healers held a strong knowledge of the inflammatory response. Yet it was the Romans that established the first written clinical definition of inflammation. According to Cornelius Celsus (circa the first century), **inflammation** was defined by four cardinal signs: *Rubor et tumor cum calore et dolore* (redness and swelling with heat and pain). In this chapter, we will approach this topic by examining the response to infection, although it is also the case that inflammation may arise in the absence of an infectious agent (think trauma and related ischemia, or response to a bee sting or toxic agent). Indeed, it

Introductory Immunology
https://doi.org/10.1016/B978-0-12-816572-0.00002-4

should be understood that inflammation may be triggered by microbial or parasitic infection. Yet any physical agent that causes direct trauma or ischemia resulting in tissue necrosis (radiation, ultraviolet light, corrosive chemicals, or burns/frostbite) also will elicit a similar cascade of responses.

Manifestation of the four signs is related to physiological changes. Redness occurs due to changes in localized blood flow (vasodilation); swelling comes from the influx of fluid and cells from blood vessels to tissue; heat is caused by increased blood arriving to areas of damage; and pain results from edema (fluid accumulation), which increases pressure on local nerves surrounding the damaged site (Table 2.1). If these four cardinal signs persist over time, a fifth component can be added to the definition, *Functio laesa*, which encompasses loss of function to the inflamed area due to destruction and scarring of living tissue.

TABLE 2.1 Effects of Fluid Exudate

	Effectors	Function
Beneficial	Entry of plasma fluid	Delivery of nutrients and oxygen
		Dilution of toxins
		Delivery of immune response mediators
		Dilution of toxins
	Entry of antibodies	Lysis of microorganisms (complement)
		Assisted phagocytosis (opsonization)
		Neutralization of toxins
	Fibrin formation	Impede movement and trap microorganisms
		Facilitate phagocytosis of foreign agents
	Entry of cells	Initiation of innate and adaptive immunity
Detrimental	Excess innate cell activation	Release of lysosomal enzymes
		Digestion or destruction of normal tissues
	Excess plasma fluid	Obstruction of ducts and lymphatics
		Vascular constriction and ischemic damage
		Pressure on nerves (pain)
Outcomes	Drainage to lymphatics	Delivery of antigens to lymph nodes
		Antigen presentation to adaptive cells

INITIATION OF THE INFLAMMATORY RESPONSE

The skin provides an effective mechanical barrier against microorganisms. Breach of that inflammatory barrier triggers an immediate cascade (Fig. 2.1). A simple tear or rip in the skin and underlying tissue causes disruption of cells and release of lipids from cellular membranes. A series of blood-borne, phospholipase enzymes are designed to transform released membrane components into signaling molecules, which subsequently can exert effects on blood vessels. Breakdown of cell membranes produces arachidonic acid, which has further effects on cyclooxygenase activity, producing **prostaglandins**; and lipoxygenase activity, which produces

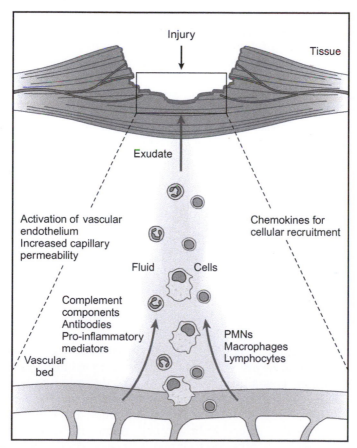

FIG. 2.1 Inflammation as a result of tissue injury. Inflammatory responses lead to vasodilation, causing erythema (redness) and increased temperature; increased capillary permeability, which allows exudate (fluid) to accumulate leading to tissue swelling (edema); and influx of cells to the site of tissue damage. Cells entering the area of injury release chemotactic factors to recruit additional cells, leading to local activation at the damaged site.

leukotrienes. Both of these agents result in vasodilation, in essence causing the endothelium lined blood barrier to become more or less "leaky."

Dilated local blood vessels, capillaries, and small venules/arterioles subsequently allow fluid (edema) to accumulate in the damaged area. If the breach in the mechanical barrier is deep enough, damaged blood vessels also directly flood the local area with leukocytes, red blood cells (RBCs), and platelets. **Plasma** (blood minus the cells) contains critical components to mediate the development of inflammation. On the positive side, the immediate consequence of tissue damage allows direct delivery of nutrients and oxygen to the affected area. Fluid entering the site also contains liver-derived factors, which are always present in circulating blood. Many of these molecules are effective at coagulation and repair, working to direct the **clotting cascade** and the production of **fibrin**. Plasma serine proteases, normally present as inactive molecules, are activated; some of these function to produce **kinins** to mediate firbrinolytic events, vascular permeability, and sensation of pain. Other proteins and enzymes, released by platelets, help to further drive and direct the initial response. Indeed, these factors initiate molecular changes that, from an immunological standpoint, are extremely beneficial. Clotting and formation of a fibrin matrix impede the movement of microorganisms in the local area; generation of breakdown products in the cascading events also attracts and activates incoming leukocytes.

THE ROLE OF ANTIBODIES IN INFLAMMATION

A major benefit of the influx of fluid is the direct delivery of relatively high levels of antibodies to the area. The circulating plasma is filled with antibodies that, as a group, have the ability to recognize just about any shape or form of antigen. Depending on prior exposure or positive vaccination status, there may also be an abundance of specific antibodies reactive to antigenic features present in destructive pathogens.

Antibody entry to the area offers multiple functions, the most critical of which resides within the end of the antibody molecule that can recognize unique shapes and forms, with a specific and unique binding site for foreign substances. This allows for binding directly to the pathogen or to pathogen-derived proteins. In effect, antibodies recognize toxins and deleterious factors; recognition results in the inhibition and neutralization of their toxic properties. Direct binding inhibits organism movement and can block attachment and adhesion to target host cells. Antibody recognition of multiple regions on the microorganism can result in a latticework structure, a precipitate, which promotes organism clearance.

The opposite end of the antibody molecule has a constant structure that confers biological function. Innate phagocytes contain surface receptors for the constant end of the antibody molecule. These receptors assist pathogen engulfment, specifically targeting foreign agent destruction using intracellular enzymes. The term for coating the organism with antibodies is **opsonization**. Targeted engulfment is referred to as **phagocytosis**.

BIOLOGICAL FUNCTIONS OF COMPLEMENT

Opsonization of organisms has another role. Simply coating the agent with antibodies targets it for attack by serum enzymes that comprise the **complement cascade** (Fig. 2.2). **Complement** is a term that refers to heat labile factors in the serum that cause immune cytolysis. As a group, complement comprises at least 50 distinct proteins that effect multiple biological functions. The complement components may be circulating in the blood, may be expressed on cell surfaces, and may even be found as intracellular-residing proteins. These components are normally inert in serum; however, a cascade of events occurs when these molecules interact with signaling agents. Activation of complement enzymes results in subsequent cleavage of proteins to yield specific polypeptide fragments with short-lived enzymatic functions related to inflammation. In essence, proteolytic degradation of the complement components drives many aspects of immunity.

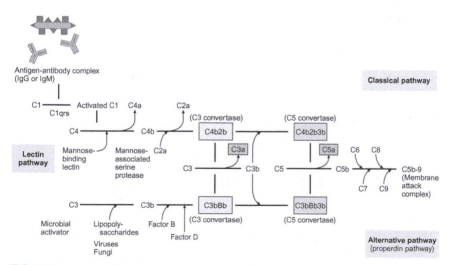

FIG. 2.2 The complement cascade. Activation of complement through the classical pathway (antigen-antibody complexes), the alternative pathway (recognition of foreign cell surfaces), or the lectin pathway (or mannose-binding pathway) promotes the activation of C3 and C5, leading to construction of the membrane attack complex.

Various pathways exist to initiate complement activation. One pathway, referred to as the **classical pathway**, utilizes antibodies as the initiating signal. The sheer action of antibodies bound to pathogenic determinants on organisms forms the basis of a physical structure. Complement components interact with the pathogen-bound antibodies. The first complement component (called C1) interacts with the constant portion of specific classes of antibody bound to the surface of the bacteria. This initiates a cascading series of reactions, whereby a complex structure is built upon the bacterial cell surface. Synthesis of this structure culminates in a pore channel, called a **membrane attack complex (MAC)**, which causes osmotic lysis of the pathogen or infected cell.

Two other pathways of complement activation, the **alternate pathway** and the **lectin pathway**, function to allow direct lysis of microorganisms in the absence of antibodies. In these enzymatic cascades, complement components bind directly to pathogens via the recognition of bacterial sugars and lipids. Similar to the events described here, the deposition of complement on the invading pathogen leads to a cascade of enzymatic reactions, culminating in pore channel assembly on the organism surface. Furthermore, breakdown products of the cascading pathway results in the production of smaller molecules, **opsonins**, which remain deposited on the pathogens. The opsonins act as molecular beacons, allowing interaction with receptors on macrophages, monocytes, and neutrophils to enhance phagocytosis and elevate mechanisms of targeted organism destruction.

Finally, a relatively new concept for intracellular functions of complement has been identified, where the complement has the ability to affect normal cell physiology. An intracellular vesicle, referred to as the **complosome**, has a strong impact on cell processes and metabolism, perhaps exerting influence on how the host cell adjusts to the presence of intracellular pathogens.

Overall, complement and its related components exert multiple biological functions that are critical components of inflammation, including activation and regulation of both innate and adaptive immune functions (Fig. 2.3). Complement will be discussed again when examining defense mechanisms against infectious agents.

ACTIVATION AND DIRECTED MIGRATION OF LEUKOCYTES

One of the most prevalent by-products of the complement enzymatic cascade is the production of molecules with the ability to call in and activate leukocytes. This chemical attraction is termed **chemotaxis**. Proteolytic degradation of complement components releases leukocyte chemotactic factors referred to as **anaphylatoxins**. For example, breakdown products of complement component C3 are chemotactic for eosinophils. Proteolytic

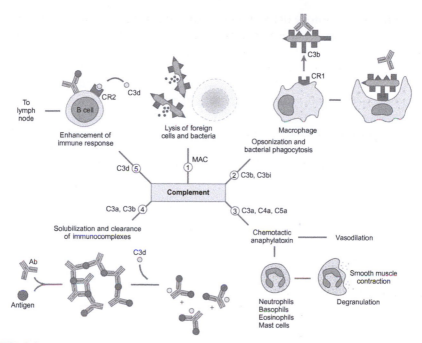

FIG. 2.3 Biologic functions of complement. Complement activation results in the formation of biologically active fragments that act as anaphylatoxins, opsonins, or chemotactic factors. Reactivity of complement with bacterial factors or with antibodies initiates the cascade. Breakdown products of the enzymatic cascade are recognized by receptors, leading to cellular activation and enhanced function. Multiple products also drive lymphocytic maturation (not shown).

digestion of component C5 produces a much more potent chemokine, attracting neutrophils, monocytes and macrophages, and eosinophils. Another major property of the enzymatic cascade is the production of factors that activate those incoming cells. For example, interaction of the breakdown components C3a, C4a, or C5a with mast cells and basophils leads to the release of histamine, serotonin, and other vasoactive amines, which further drives vascular permeability. One can readily see that by-products of the complement cascade can directly influence local inflammatory responses.

PATHOGEN RECOGNITION AND CYTOKINE SIGNALING

Phagocytes bear several unique receptors that recognize microbial components, bind bacterial carbohydrates, and induce phagocytosis. Recognition can occur directly through **mannose receptors**, **scavenger receptors**, or **Toll-like receptors** (**TLRs**), receptors that detect common pathogenic motifs (**pathogen-associated molecular patterns**; **PAMPs**). Recognition through any of these receptors represents a "danger" signal that initiates a proinflammatory response (Fig. 2.4); triggering of these cell surface sensors activates gene regulation. In addition, internal protein complexes, such as the **inflammasome**, further the response to the recognition event via internal processing of inactive forms of proinflammatory molecules to produce mature and active mediators. The result is the release of small chemical mediators called **cytokines**. Cytokines represent cell-derived secreted mediators that allow cells to communicate with each other. They regulate development and behavior of immune effector cells and facilitate cross-talk at low concentrations (10^{-10}–10^{-15} M). They are short-lived and bind to cell surface receptors. They may act alone or in concert with one another synergistically. They are considered pleiotropic because they exert multiple actions on multiple different cell types, with overlapping and redundant functions. Another common name for these molecules is **lymphokines**, which are basically cytokines produced by cells of the lymphoid lineage. Finally, many cytokines also function as growth factors for specific cell subsets (Table 2.2).

It was mentioned earlier in this chapter that trauma also may serve as a trigger for inflammation. In addition to the PAMP molecules described here, a unique subset of **danger-associated molecular patterns** (**DAMPs**), can trigger similar responses. In this case, innate immune recognition of host intracellular molecules released by injured tissue can jump-start the immune response. Multiple DAMP components have been identified, including those released during cell death from the nucleus, the cytosol, and the mitrochondria.

FEEDBACK AND ADAPTATION FROM A DISTANCE

There is considerable activity between cells at the local site of the inflammatory response. Indeed, we will see in later chapters that the direct interaction of cells that initiate the proinflammatory response with incoming lymphocytes triggers adaptive cell activity. Cell-to-cell contact gives the advantage of directly delivering cytokines and mediators, allowing functional lymphocyte development in the immediate area to target productive immune function. However, there is also a great need for systemic communication that can mediate responses across great distances to direct cells and organs located distant from the site of tissue damage.

A special subclass of chemical mediators responsible for attracting cells to the area of inflammation is called **chemokines**. Chemokines assist in leukocyte migration into tissue (**diapedesis**). As a class, they are small polypeptides synthesized by a wide variety of cell types, all of which

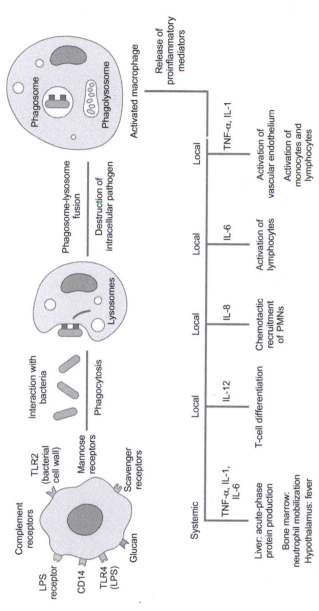

FIG. 2.4 Important cytokines secreted by macrophages. Recognition of bacteria and bacterial products induced cytokine that regulate both local and systemic events involved in the inflammatory response to contain infections. *IL*, interleukin; *TNF-α*, tumor necrosis factor-alpha; *LPS*, lipopolysaccharide; *CD14*, LPS and/or lipoteichoic acid receptor; *PMN*, polymorphonuclear cell; *TLR*, toll-like receptor.

TABLE 2.2   Selected Cytokines and Functions

Cytokine	Cell source	Cell target	Primary effects
IL-1α, IL-1β	Monocytes	T cells	Costimulatory molecule
	Macrophages	B cells	Activation (inflammation)
	Fibroblasts	Endothelial cells	Adhesion, fever
	Epithelial cells		Acute phase reactants
	Endothelial cells		
IL-2	T_H1 cells	T cells	Growth and activation
	NK cells	B cells	
		Monocytes	
IL-3	T cells	Bone marrow hematopoietic precursors	Growth and differentiation
	NK cells		
	Mast cells		
IL-4	T_H2 cells	Naive T cells	Differentiation to T_H2 cell
	Mast cells	T cells	Growth and activation
		B cells	Isotype switching to IgE
IL-5	T_H2 cells	B cells	Growth and activation
	Mast cells	Eosinophils	
IL-6	T cells	T cells	Costimulatory molecule
	Macrophages	B cells	Adhesion, fever
	Fibroblasts		Acute phase reactants
IL-8	Macrophages	Neutrophils	Chemotaxis and activation
	Epithelial cells		
	Platelets		
IL-10	T_H2 cells	Macrophages	Inhibits T_H1 development
		T cells	
IL-12	Macrophages; NK cells	Naive T cells	Differentiation into a T_H1 cell

TABLE 2.2   Selected Cytokines and Functions—cont'd

Cytokine	Cell source	Cell target	Primary effects
IL-13	T cells	B cell	IgE switching
			B-cell growth
IL-17	T cells	Neutrophils, inflammatory cells	Inflammatory regulation, chronic inflammation
IL-18	Macrophages	T cells, NK cells	Increase IFN-γ production
IFN-α, IFN-β	T cells	Monocytes	Activation
	NK cells	Endothelial cells	Increased class I and II MHC
		Macrophages	
IFN-γ	T_H1 cells	Monocytes	Activation
	NK cells	Endothelial cells	Increased class I and II MHC
		Macrophages	
TGF-β	T cells	T cells	Inhibits activation and growth
	Macrophages	Macrophages	
GM-CSF	T cells	Bone marrow progenitors	Growth and differentiation
	Macrophages		
	Endothelial cells		
	Fibroblasts		
TNF-α	Macrophages	T cells, B cells, endothelial cells	Costimulatory molecule
	T cells		Activation (inflammation)
			Adhesion, fever, acute phase reactants

IL, interleukin; *IFN*, interferon; *TGF*, transforming growth factor; *GM-CSF*, granulocyte-macrophage colony stimulating factor; *TNF*, tumor necrosis factor.

act through receptors that are members of the G-protein-coupled signal-transducing family. All chemokines are related in an amino acid sequence. Their receptors are integral membrane proteins characterized by a common physical shape containing seven membrane-spanning helices.

TABLE 2.3 Partial List of Chemokines Produced by Monocytes and Macrophages

Chemokine class	Chemokine	Cells affected
CC	MCP-1 (CCL2)	Monocytes, NK cells, T cells, dendritic cells
	MIP-1 alpha (CCL3)	Monocytes, NK cells, T cells, dendritic cells
	MIP-1 beta (CCL4)	Monocytes, NK cells, T cells, dendritic cells
	Rantes (CCL5)	Eosinophils, basophils, monocytes, dendritic cells, T cells
	Eotaxin (CCL11)	T cells, eosinophils
CXC	IL-8 (CXCL8)	Neutrophils (naive T cells)
	IP-10 (CXCL10)	Monocytes, NK cells, T cells

Chemokines fall mainly into two distinct groups. CC chemokines have two adjacent cysteine residues (hence the name CC). CXC chemokines have an amino acid between two cysteine residues. Each chemokine reacts with one or more receptors and can affect multiple cell types (Table 2.3).

As the focal point of inflammation spreads, an amplification of cognate component signals occurs. Endothelial cells (cells that make up blood vessel walls) and incoming leukocytes generate and release additional prostaglandins, leukotrienes, platelet-activating factors, and enzymes. Factors draining back to the blood supply circulate to organs and initiate responses from tissues distant to the region of inflammation. For example, signals received by the hypothalamus can trigger increased body temperature (fever), while signals to the liver trigger the synthesis of **pentraxins** (a class of innate pattern recognition molecules) and **acute phase proteins**, including clotting factors and complement components. The liver also produces the pentraxin **C-reactive protein** (**CRP**), which enables increased activation of complement during times of infection.

The acute phase proteins, in combination with chemical mediators and cytokines released from cells at the site of infection, promote the release of neutrophils from bone marrow. Related chemokines released at the site of inflammation also attract myeloid-derived granulocytes, which specifically function to combat infectious agents. This culminates in the recruitment of neutrophils and **polymorphonuclear cells** (**PMNs**) to the site of tissue damage. Arriving neutrophils become engaged in the inflammatory process; neutrophils contain primary and secondary granules containing specific proteases to kill microorganisms. Activated neutrophils express high levels of antibodies and complement receptors, allowing increased phagocytosis of invading organisms. Activation of neutrophils leads to a respiratory burst, producing reactive oxygen and nitrogen intermediates, which directly kill invading pathogens. Neutrophilic myeloperoxidases also function to destroy pathogenic invaders.

PATHOLOGICAL CONSEQUENCES OF THE INFLAMMATORY RESPONSE

The initiation of the proinflammatory responses is extremely potent to limit the spread of microorganisms; however, full clearance of the infectious agent typically requires adaptive immune components. Fig. 2.5 schematically reviews the stimuli, cell responses, and pathological consequences that are involved in moving from acute to chronic inflammation. The productivity and success of the acute inflammatory response lie in the ability of external or internal agents to activate innate responses that lead to the production of proinflammatory mediators. Assistance from environmental cues (released clotting factors, kininogens, and complement) is critical to the preparation of vascular beds to allow cellular influx.

Activation of local response is followed by activity of the lymphatics to functionally drain fluids, cells, and debris to nearby lymph nodes, where phagocytic antigen presenting cells "show" digested pieces of the foreign protein to cells of the adaptive immune response. In essence, the next step

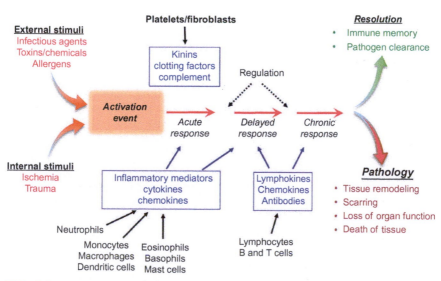

FIG. 2.5 Biological elements and consequences of inflammation. Multiple parameters drive the inflammatory response, leading to the development of pathological consequences and clinical outcomes.

is a "priming" of lymphocytes for specific activation. The activation of lymphocytes leads to the release of antibodies, cytokines, and growth factors specifically targeting the causative agent of the inflammatory response. As a clinical manifestation, the triggering of adaptive components leads to lymph node swelling, a simple sign that the immune system is undergoing high activity. We will see in the next few chapters how the activation of lymphocytes leads to directed amplification of the adaptive response, which is required for continued protection against foreign agents, and either resolution of inflammation or chronic activity. We also will see that when the inflammatory response remains persistent, the chronic inflammation can lead to permanent tissue damage.

SUMMARY

- A coordinated effort of cells and blood components is required to elicit the inflammatory response. Inflammation manifests as redness, swelling, heat, pain, and loss of function (rubor, tumor, calor, dolor, and functio laesa).
- The inflammatory process is initiated and controlled via chemical mediators, resulting in an influx and activation of inflammatory cells.
- Antibodies and complement components are critical to the initiation of innate immune functions, working independently and in concert to eliminate microorganisms. By-products of the complement enzymatic cascade function to both heighten and direct later responses.
- Recognition of foreign agents leads to the release of cytokines and chemokines to function locally to attract granulocytes and leukocytes to the site of inflammation. These factors also work systemically in the release of acute phase proteins to further drive protective responses.
- The response culminates in the draining of fluid, cells, and antigens to local lymph nodes, where maturation and activation of adaptive lymphocytes occurs.

3

The B Lymphocyte: Antibodies and How They Function☆

CHAPTER FOCUS

To discuss factors involved in generation and establishment of the humoral response. The structure of the immunoglobulin will be presented as a way to understand the biological functions balanced with the molecule's ability to recognize unique antigenic determinants. A discussion will examine each immunoglobulin isotype to detail how structural features confer biological properties, mediated in part through interactions with receptors on effector cell subsets. B-lymphocyte development will be addressed as it relates to producing cells with the capability to synthesize immunoglobulins, examining concepts of gene rearrangement to allow the generation of unique antigen-binding structures.

B LYMPHOCYTES PRODUCE ANTIBODIES

In 1965, pathologist David Glick reported that removal of the bursa of Fabricius, a hematopoietic organ located near the cloaca of chickens, resulted in a significant decrease in circulating antibodies. Orthologous lymphocytes in humans were determined to develop in the bone marrow. It is now understood that the B subset of lymphocytes is responsible for **humoral immunity**, defined by their expression of antibody molecules.

☆This chapter is contributed in collaboration with Keri C. Smith, PhD.

Introductory Immunology
https://doi.org/10.1016/B978-0-12-816572-0.00003-6

31

STRUCTURAL CHARACTERISTICS
OF IMMUNOGLOBULINS

Immunogloblulins, interchangeably referred to as **antibodies**, share a common structure that allows them to bind to a nearly limitless number of specific antigens (including proteins, carbohydrates, glycoproteins, polysaccharides, nucleic acids, and lipids). Their structure confers multiple cellular processes, mediated by variable and constant region domains.

The anatomy of the immunoglobulin in its monomeric form may be described as a Y-shaped glycoprotein consisting of two identical heavy chains of 55 kDa, paired with two identical light chains of 25 kDa (Fig. 3.1). These chains are held together by one or more interchain disulfide bonds; enzymatic cleavage of the disulfide bonds results in two fragments—a homogenous portion that could be crystallized (fragment, crystalline; **Fc**) and another fragment that could bind antigen (**Fab2**). These regions denote the effector antigen-binding functions of the antibody molecule. Within the heavy chain at the junction of the Fab$_2$ and Fc regions are short amino acid

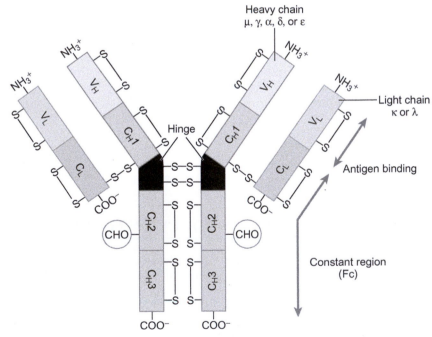

FIG. 3.1   Anatomy of the immunoglobulin. The basic structure of the antibody contains heavy chains and light chains, showing intradisulfide and interdisulfide bonds and the characteristic hinge region. The interactions between variable domains constitute the antigen-binding domain, while the constant regions confer specific biological properties of the molecule.

sequences that are rich in proline and cysteine. These are **hinge regions,** which confer flexibility for optimized binding with antigen.

Each heavy chain pairs with one of two varieties of light chains, the kamma or lambda light chain, to comprise the Fab_2 portion of the antibody. Either chain variety, but not both, may be used in an individual antibody molecule—approximately 60% of human antibodies utilize the kamma chain, and the remainder contain the lambda chain. In humans, five unique heavy chains are defined by differences in amino acid sequences. They are labeled according to the Greek letter designations alpha gamma, mu, epsilon, and delta. The combination of one specific heavy chain with one light chain results in a specific antibody **isotype** now referred to as **immunoglobulin A, immunoglobulin G, immunoglobulin M, immunoglobulin E,** or **immunoglobulin D (IgA, IgG, IgM, IgE,** or **IgD).** The properties of these specific isotypes are summarized in Table 3.1.

As with other members of the Ig superfamily (which include T-cell receptors discussed in Chapter 4), each chain shares a similar tertiary structure, with characteristic "immunoglobulin folds" consisting of "beta-barrel" antiparallel beta-pleated sheets connected by loops of variable length stabilized by at least one disulfide bond. The overall structures of the specific subclasses of immunoglobulin folds confer biologic functions.

The **variable domain**, comprised of the light-chain and heavy-chain heterodimers, is responsible for antigen binding. Each individual antibody clone expresses a unique variable domain. Within the 100–110 amino acids that make up this domain are regions of extreme sequence variability, called **hypervariable regions**. Each light chain and each heavy chain expresses three of these unique regions, which make up the **complementarity-determining region (CDR)**, so called because its protein structure complements antigen binding. The Fab portion contains two identical variable domains; therefore, each complete molecule is capable of simultaneously binding two identical antigens. The regions between them are less variable framework regions which form the structural support to allow antigen contact. Importantly, antigen binding occurs through noncovalent means (e.g., van der Waals forces, hydrogen bonding, hydrophobic bonding, and electrostatic interactions). Hence, the three-dimensional (3D) structure formed by the combination of the protein chains is essential for proper antigen binding.

The **constant domains** provide structural stability. Each light chain contains one constant domain, while heavy chains have three to four. These domains associate to form the Fc region of the antibody which, depending on isotype, mediates various downstream effects, including complement binding, opsonization, and phagocyte activation. Properties specific for the Fc region of each isotype are described in detail next.

TABLE 3.1 Classes of Antibody Isotypes and Their Functional Properties

Isotype	Immunoglobulin class				
	IgM	IgD	IgG	IgE	IgA
Structure	Pentamer	Monomer	Monomer	Monomer	Monomer, dimer
Heavy-chain designation	Mu	Delta	Gamma	Epsilon	Alpha
Molecular weight (kDa)	970	184	146–165	188	160×2
Serum concentration (mg/mL)	1.5	0.03	0.5–10.0	<0.0001	0.5–3.0
Serum half-life (days)	5–10	3	7–23	2.5	6
J chain	Yes	No	No	No	Yes
Complement activation	Strong	No	Yes, except IgG4	No	No
Bacterial toxin neutralization	Yes	No	Yes	No	Yes
Antiviral activity	No	No	Yes	No	Yes
Binding to mast cells and basophils	No	No	No	Yes	No
Additional properties	Effective agglutinator of particulate antigens, bacterial opsonization	Found on the surface of mature B cells, signaling via cytoplasmic tail	Antibody-dependent cell cytotoxicity	Mediation of allergic response, effective against parasitic worms	Monomer in secretory fluid, active as dimer on epithelial surfaces

IMMUNOGLOBULIN NOMENCLATURE

Although all immunoglobulins are remarkably similar in structure, changes in amino acid sequences can led to obvious differences in functions, structure, and antigen specificity (Fig. 3.2). These differences are denoted by several different terms: The most structurally diverse are the **antibody isotypes** which are defined by their expression of different heavy chains (alpha, gamma, mu, delta, or epilson). The IgG and IgA isotypes are subdivided further into specific **subclasses**. These subclasses

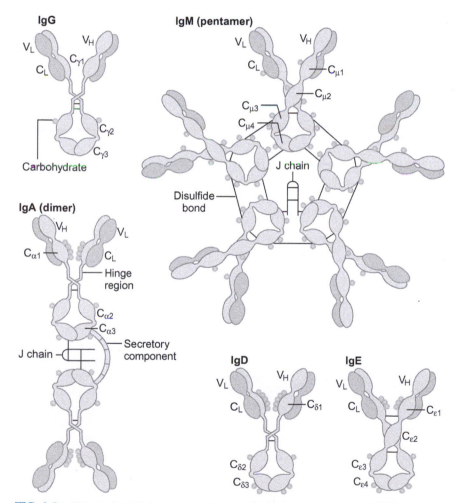

FIG. 3.2 Classes of antibody isotypes. Shown here are the five classes of antibody isotypes, representing monomeric IgG, IgD, and IgE, pentameric IgM, and dimeric IgA with secretory components.

share 90% amino acid homology, but the differences in protein folding and disulfide bonding considerably affect their biologic properties. Within the same isotype and subclass, individuals can express unique allelic variants, called **allotypes**, of light and heavy chains. Allotypes generally consist of one amino acid substitution at a specific site. They are inherited in a codominant autosomal Mendelian pattern and have been used for forensic purposes. Large-scale studies associate certain allotypes with disease progression or with development of autoimmunity. Finally, the hypervariable regions of the light and heavy chains represent a signature unique to a single antibody, also known as an **idiotype**.

BIOLOGIC PROPERTIES OF ANTIBODY ISOTYPES

IgD

The IgD immunoglobulin is expressed primarily as a membrane-bound monomeric form on naive B cells. Only small amounts of soluble IgD can be detected in serum. It has a molecular weight of 180 kDa and contains a relatively flexible hinge region. The function of this antibody as a surface receptor is not entirely understood, though some evidence suggests that it may facilitate the binding of antigen and play a role in affinity maturation. Expression of IgD is lost after responding to antigen because of class-switch recombination events.

IgM

IgM can also be expressed in a membrane-bound form on mature, naive B cells. Its expression is lost following class-switch recombination. However, due to alternative RNA-processing mechanisms, IgM also can be produced in a secreted form by **plasma cells**, the activated form of B lymphocytes. IgM plays a very important role in primary antibacterial immune response and is the first antibody isotype secreted during an immune response. Elevated levels in adults indicate recent antigenic exposure.

The secreted, pentameric form of IgM is very large—900 kDa in size. Each monomer interacts with a small joining polypeptide, called a *J chain*, to form a pentameric molecule consisting of five monomers. Because each original monomer has two antigen-binding sites, each pentameric complex is able to bind to 10 identical antigens simultaneously. Because of this, IgM has a greater **functional avidity** than monomeric IgG. However, due to lack of flexibility in its hinge regions and steric hindrance that occurs on larger antigens, its effective valency is lowered. IgM is secreted before affinity maturation takes place and generally has a lower affinity for antigen than IgG or IgA.

This high functional avidity due to the cross-linking of multiple antigens can be best demonstrated by the ability of IgM to aggregate or agglutinate particulate antigens such as bacteria or red blood cells. As such, IgM is very effective at neutralizing bacterial antigens early in the immune response. A clinical manifestation of IgM binding can be observed in assays used to determine blood type. Naturally occurring IgM antibodies reactive with red blood cell antigens of the ABO series are referred to as **isohemagglutinins**. Precipitation reactions occur when a specific isohemagluttinin is present in serum that comes in contact with the appropriate blood group antigens. The pentameric structure of IgM is particularly effective at mediating complement activity. Indeed, IgM is the most efficient isotype on a mole:mole basis for complement activation of the classical pathway. The majority of IgM is located in intravascular spaces; its large size precludes traversing the maternal/placental barrier.

IgG

Monomeric-secreted IgG is the smallest of the isotypes, with a heavy chain consisting of three constant domains. IgG is primarily secreted in the serum, where it can account for up to 15% of total protein in the human body. Due to its small size, IgG is capable of crossing endothelial cell barriers and is approximately equally distributed between extravascular and intravascular spaces. The small size of IgG also allows it to cross the placenta; maternal-derived IgG provides passive immunity to the neonate for up to 4 months after birth.

IgG is the primary immunoglobulin present in blood and extravascular spaces. It can effectively bind to bacterial toxins, viral receptors, or bacterial adhesions to act in a **neutralizing** capacity to prevent pathogen binding to host cell surfaces. Binding of IgG to antigens also can result in downstream activation of antigen-presenting cells via coordinated interaction with specific surface receptors (FcγR). This process is known as **opsinization**, derived from the Greek phrase *opsonin*, meaning "to prepare for eating," as it allows for targeted uptake of antigen in a phagosome. The Fc portion of IgG plays an important role in **antibody-dependent cell-mediated cytotoxicity** (**ADCC**). In this instance, binding of antigen-IgG complex to FcγR expressed by natural killer (NK) cells helps to guide destructive granules to a target cell. IgG also plays an important role in the **activation of the complement cascade**; specifically, binding of multiple IgGs to antigen leads to the initiation of classical complement activity.

Differences in the amino acid sequences, as well as in the number and location of disulfide bonds in the gamma chain, result in four different IgG subclasses. These subclasses, numbered according to prevalence in human

TABLE 3.2 Unique Biologic Properties of the Human IgG Subclasses

	IgG1	IgG2	IgG3	IgG4
Occurrence (% of total IgG)	70	20	7	3
Half-life (days)	23	23	7	23
Complement binding	+	+	Strong	No
Placental passage	++	±	++	++
Receptor binding to monocytes	Strong	+	Strong	±

serum, differ (Table 3.2). The IgG response to specific antigens is T-cell dependent, meaning that it requires T helper cell-derived cytokines and physical interactions for subsequent class switching. IgG appears in serum within 2 weeks of primary antigen exposure, and it is very rapidly produced at even higher levels following a second antigen "boost." This anamnestic response is the basis of vaccination strategies that result in long-term immune protection against pathogens (see Chapter 9).

IgA

IgA is best known as the hallmark immunoglobulin of the mucosal immune response, and the dimeric, secretory form is found in seromucous secretions, including saliva, tears, nasal fluids, sweat, colostrum, and secretions of the genitourinary and gastrointestinal tracts. The alpha chain consists of three constant regions, and like the mu chain, it also has a cysteine-containing tailpiece that allows for the association of the polypeptide J chain and the formation of the dimeric structure. Of note, IgA is secreted on the basal side of mucosal surfaces by plasma cells in lamina propria. The assembled dimer binds via a polymeric Ig receptor on the basal surface of the mucosal epithelial cell, followed by receptor-mediated endocytosis and transport of IgA to its apical surface. The polymeric Ig receptor is partially cleaved from the IgA dimer, and the remaining attached portion is referred to as **secretory component**. Secretory component functions to bind the IgA to the mucous surface and protect the antibody against cleavage from digestive enzymes.

IgA is produced in very large quantities (as high as 3 g/day in the intestine) and can neutralize bacteria. It also has a high viricidal activity; its dimeric form can efficiently agglutinate viruses. IgA is not capable of activating the complement pathway, but it has been demonstrated to have bactericidal activity for Gram-negative organisms, although this requires the presence of endogenous lysozyme. IgA can be found in serum, where it exists as a 160-kDa monomer that neutralizes antigen in the blood and extravascular space. Finally, the transfer of IgA from mothers' milk to the intestinal tract of the infant provides passive immunization against pathogens in the neonate. IgA can be secreted as one of two subclasses, which differ in the length of hinge region and in the number of disulfide bonds. In mucosal secretions, the balance between the IgA1 and IgA2 subclasses is approximately 1:1; however, the serum form of IgA is almost exclusively monomeric IgA1.

IgE

IgE molecules are normally only present in trace amounts in serum and have a very short half-life. IgE is secreted in a monomeric form and is a highly potent activator of mast cells—very little IgE is required to induce a very obvious downstream response. By itself, IgE does not activate complement or neutralize bacteria or viruses. Instead, it is secreted by plasma cells in response to antigen and is then bound via its Fc portion to FcεR expressed on mast cells. In this form, it can be retained on the cell surface in extravascular space for weeks to months. Mast cells may be considered a cellular form of an "armed bomb"; they contain granules loaded with histamine, heparin, and leukotrienes, and can rapidly synthesize and secrete prostaglandins and cytokines. Upon secondary encounter with antigen, cross-linking of two or more of the "preloaded" IgE-FcεR complexes triggers mast cell degranulation.

IgE is necessary for human health in the face of infection with specific parasites, notably helminths. Antigen stimulates IgE bound to mast cells; the degranulation of released mediators increases vascular permeability and local inflammation. This results in the recruitment of eosinophils from the blood to the site of the parasitic infection. Eosinophils also can bind to IgE attached to the surface of the parasite, and then release the contents of their granules to destroy the worm via an ADCC-type mechanism. IgE is an important part of the first line of defense against pathogens that enter the body across epithelial barriers. A deleterious effect of IgE can occur when it binds to normally innocuous antigens such as pollen, triggering mast cell degranulation associated with allergic responses.

KINETICS OF ANTIBODY RESPONSE

Antibody isotypes are found in higher concentrations during an active immune response. Typical immune responses in humans, as defined by the detection of antibodies in serum specific for injected antigen, have four phases, the length of which depends upon whether the response is primary or secondary (memory, anamnestic). In the primary response, B lymphocytes need time to receive help from T cells. This period can last up to 2 weeks.

IgM production occurs first (sometimes followed by IgG class switching). Next follows the exponential phase, in which the concentration of antibody in serum increases exponentially, which corresponds to the development of plasma cells secreting class-switched antibody. A steady state of antibody secretion and degradation is then maintained, and when the antigenic threat has been dealt with, the response enters the declining phase. Plasma cells die, but memory B cells remain.

Upon restimulation, these memory cells generate the anamnestic immune response, characterized by a very short lag phase, a greater production of antibody in the exponential phase, and a longer period of steady-state antibody maintenance. Class-switched antibodies, including IgG, IgA, and IgE, are more likely to appear in this secondary response, and affinity maturation often occurs in the rapidly dividing plasma cell population. The capacity to generate a secondary response may persist for many years if not decades.

MEMBRANE-BOUND IMMUNOGLOBULIN

Heavy chains may contain a transmembrane domain that allows for Ig to be expressed on the surface of B cells. They allow for antigen-specific binding and subsequent activation of B lymphocytes. Importantly, antigen binding by the antibody is not sufficient to induce a cellular activation signal. Further association of the Ig-heavy chains with two transmembrane invariant protein chains, Igα and Igβ, is required to generate a cellular signal. Igα and Igβ contain cytoplasmic immune-receptor tyrosine-based activation motifs (ITAMs) that are phosphorylated following binding of the membrane-bound Ig to multivalent antigen.

DEVELOPMENT OF B CELLS

Hematopoietic stem cells in the bone marrow give rise to multipotent cells, resulting in the development of a common lymphoid progenitor. This cell can produce the earliest committed B cell—the pro-B cell. The developmental stages of the B cells correspond to gene rearrangement events, ultimately culminating in the formation of cells that express a vast number of unique receptors for antigen.

GENE RECOMBINATION

The antigen-binding capacity of the B-cell receptor (BCR) is generated before the cell is exposed to antigen. Rearrangement of multiple genes is necessary to produce the high diversity needed for a broad immune response (Fig. 3.3). Recombination of both heavy- and light chain genes is accomplished via the actions of the **V(D)J recombinase** enzyme complex. Germline DNA which encodes the heavy-chain genes, contains three

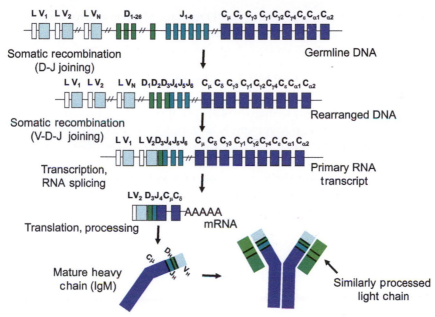

FIG. 3.3 Genetic organization and recombination events. Antibody diversity is generated by DNA recombination events that randomly fuse variable, diversity, and joining regions. The recombination is accomplished in a defined order by the recombinase enzymes RAG-1 and RAG-2. The first events culminate in transcribing messenger RNA (mRNA) coding for IgM and IgD; differential translation determines whether mature polypeptide will be IgM or IgD. During plasma cell differentiation, the isotype may be changed where the same variable region is recombined with a different constant region sequence (not shown). L, leader sequence; V, variable; D, diversity; J, joining; C, constant.

different regions, V_H (variable), D_H (diversity), and J_H (joining), which contribute to production of the **variable domain**. A V(D)J recombinase mediates joining of gene segments by splicing unused segments out of the genome. The rearranged gene segments are then brought together and juxtaposed with the segments coding for the constant region of the heavy chain. The combining of genes is a "sloppy" process. Extra nucleotides are inserted by the enzyme deoxynucleotidyl transferase to fill spaces. This process, called *N region diversification*, expands the potential pool of BCRs generated by effectively creating new codons between junctions. Rearrangment of the light-chain variable domain occurs in the same fashion, though the germline DNA lacks D-gene segments. The process of V(D)J rearrangement culminates in the generation of three CDRs in each variable region. Interestingly, two of the three CDRs are hardwired into the V gene segment—they are dependent on the V segment selected during rearrangement. The third CDR consists of the junction of the V, J, and (in heavy chain) D segments, and thus has higher variability. The combinatorial diversity that results from the rearrangement of the heavy chain (50 V_H genes × 26 D_H genes × 6 J_H genes) yields >7500 possible V(D)J loci combinations. The light-chain germline DNA sequences, which lack D regions, result in approximately 200 and 160 different VJ combinations, respectively.

DEVELOPMENT AND SELECTION OF MATURE B CELLS

The pro-B cell begins to mature with V(D)J recombination of the heavy-chain genes. If this rearrangement is accomplished successfully, the complete Ig-heavy chain is expressed by the cell as a *pre-B receptor*. Production of a functional heavy chain inhibits further rearrangement of the heavy-chain genes (enforcing allelic exclusion). These large pre-B cells then proceed with recombination of their light-chain genes, after which the cell expresses a complete IgM molecule on the cell surface and is identified as an immature B cell.

The immature B cell is then tested for autoreactivity. If the IgM expressed on the cell surface reacts with multivalent self-molecules, such as major histocompatibility complex (MHC) expressed on stromal cells, the cell is considered defective and must be removed from the cell repertoire. This is accomplished by triggering apoptosis in the self-reactive B cells (clonal deletion). Reactivity with self also can trigger receptor editing, where the V(D)J recombinase remains active and continues to rearrange light-chain DNA until a nonautoreactive BCR is produced. If an immature B cell shows no reactivity with self-antigen, it migrates from the bone marrow. The mature B cell takes up residence in peripheral lymph tissues.

Some B cells weakly reactive to self-molecules may be allowed to exit the bone marrow. Under normal circumstances, these cells remain ignorant of their self-antigen, but under certain conditions (e.g., inflammation), they may become active. Such sleeper B cells are thought to play a role in some autoimmune diseases. Together, the clonal deletion, receptor editing, and anergy induced within bone marrow form the mechanisms of central tolerance.

ACTIVATION AND DIFFERENTIATION OF B CELLS

Mature B cells in the periphery that have not yet responded to antigen are considered naive. Response to antigen depends on the type of epitope recognized by the variable regions, as well as signaling through nearby membrane coreceptors (CD19 and CD21). BCRs specific for protein antigens require help from CD4+ T cells that recognize related antigenic determinants (as discussed further in Chapter 4). Secretion of cytokines by T cells then drives B-cell maturation and proliferation. This process results in the generation of plasma cells—fully mature, class-switched B cells that secrete antibodies and no longer express membrane-bound immunoglobulins.

Importantly, interaction of B cells with T cells drives the process of iso-type switching, also known as *class switching*. **Isotype switching** is the process by which the antigen specificity of the antibody (located within the variable domains) remains the same, but the mu- or delta-heavy-chain gene locus is excised and replaced with another heavy-chain gene-constant domain. Specific cytokine signals determine which heavy chain is substituted, and once the intervening DNA between recombination sites is deleted, the switch is irreversible. Interactions of B and T cells with antigen also can result in multiple rounds of mutation and selection for higher-affinity BCR in the germinal centers of peripheral lymph tissues. Somatic hypermutation can result in amino acid replacements in the V regions of H and L chains, and the selection of cells with more tightly binding BCR to the antigen occurs, commonly known as **affinity maturation**.

While most of the B cells that proliferate in the germinal centers differentiate into plasma cells, a distinct population may follow a different path to become memory B cells. Upon reencounter with antigen, they can rapidly divide and produce high-affinity secreted antibodies. Repeated doses of antigen, such as those that occur during immunization boosters, result in the production of increased numbers of memory cells.

Finally, it should be noted that subpopulations of B cells exist. The classical B cells discussed thus far in this chapter are referred to as **conventional B-2 cells**. The B-2 cells comprise the majority of the B cell population and have extensive diversity in their antibody repertoire. There exists a second population of B cells, called **B-1 cells**, in which limited antibody diversity occurs. The B-1 population is primarily localized in the peritoneal and pleural cavities; this population secretes mainly IgM isotype antibodies and has little requirement for T-cell helper activity. This minor population of B cells may represent an innate B-cell lineage, with primitive activity to monitor active infections. The naive circulating IgM present in blood is likely produced by this B-1 population. In a similar manner, a population of B cells found in the white pulp of the spleen, referred to as **marginal zone B cells**, also represents limited antibody diversity. These marginal zone B cells primarily secrete IgM and are thought to react to common bacterial carbohydrates, independent of T-cell help.

SUMMARY

- B lymphocytes produce immunoglobulins (antibodies) with specific biological functions that confer humoral immunity.
- During B-cell development, rearrangement of the germline DNA generates antigen-binding diversity. Recombination events occur prior to the detection of antigen.

- There are five sources of antibody diversity: (1) the presence of multiple V-gene segments; (2) a combinatorial diversity of random recombination of V, D, and *J* segments; (3) junctional and insertional diversity altering V-D and D-*J* junctions; (4) coexpression of H- and L-chain pairs; and (5) somatic hypermutation.
- Each immunoglobulin isotype confers unique biological properties, mediated through interactions with antigens, as well as through receptors on effector cells. Isotype switching occurs after antigenic stimulation and requires T cell-produced cytokines.
- Antibodies themselves are neutral; they can be either protective or destructive, depending on multiple immune parameters.

T Lymphocytes: Ringleaders of Adaptive Immune Function

CHAPTER FOCUS

To examine T lymphocytes as regulators of adaptive immune function. As such, they function as the primary effectors for cell-mediated immunity. An illustrative discussion will detail cellular development, moving from bone marrow precursors undergoing antigen receptor gene rearrangement through thymic selection and subsequent maturation events that permit antigen recognition. The process of antigen presentation by major histocompatability molecules will be covered, followed by an analysis of effector functions that allow T cells to regulate immunosurveillance toward infectious assault.

T LYMPHOCYTES: SPECIFIC AND LONG-LASTING IMMUNITY

The immune system must be able to recognize a large pool of antigens while maintaining the ability to distinguish between foreign material and the self. The **T lymphocytes (T cells)** confer response specificity using surface antigen receptors to facilitate the recognition of foreign material. These adaptive lymphocytes can respond to a wide array of antigens; however, the speed of response is slower than innate functions due to intermediate steps required from the time of infection to the mounting of a protective response. In essence, the T cell acts as a master ringleader of cell-mediated immunity, a task accomplished by direct cellular contact or via secreted cytokine factors.

THE T CELL RECEPTOR

The **T cell receptor (TCR)** is a transmembrane heterodimer composed of two disulfide-linked polypeptide chains (Fig. 4.1). Each lymphocyte carries a TCR of only a single specificity. T lymphocytes can be stimulated by antigens to give rise to progeny with identical antigenic specificity. The vast majority of T lymphocytes express alpha (α) and beta (β) chains on their surface. Cells that express gamma (γ) and delta (δ) chains comprise only 5% of the normal circulating T-cell population in healthy adults. Each chain (α, β, γ, or δ) represents a distinct protein with approximate molecular weight of 45 kDa. An individual T cell can express either an $\alpha\beta$ or a $\gamma\delta$ heterodimer as its receptor, but never both. The TCR is always expressed with the associated **CD3 complex**, comprised of multiple independently expressed units, required for signal transduction once the presented antigen is encountered.

T-CELL DEVELOPMENT

All cells of the lymphoid lineage are derived from the common lymphoid progenitor cell, which differentiates from bone marrow hematopoietic stem cells. T-cell precursors migrate to the thymus where they develop and undergo thymic selection to eliminate autoreactive cells (Fig. 4.2). As

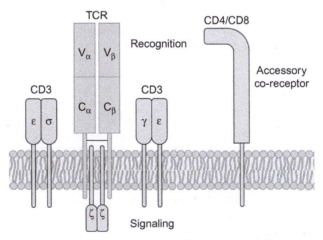

FIG. 4.1 Structure of the TCR complex. The TCR is comprised of two covalently linked polypeptide chains. The predominant antigen-binding chains, α and β, are shown. Both transmembrane peptides exhibit a variable (antigen recognition) and a constant external domain connected via a disulfide bond. The TCR is always expressed with the associated CD3 complex, required for signal transduction. T-helper cells (T_H) express CD4, which is required for interaction with APCs, whereas CTLs express the CD8 coreceptor molecule.

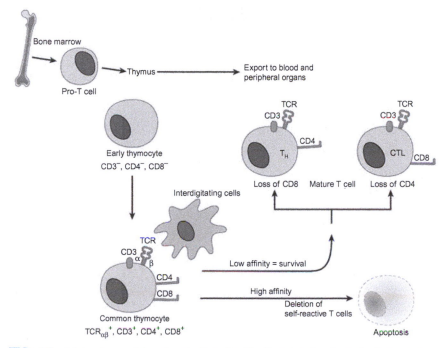

FIG. 4.2 Main stages of thymic selection. Pro-T cells migrating from the bone marrow enter the thymus, where they express rearranged TCR, CD3, CD4, and CD8 proteins. Positive selection occurs to eliminate self-reactive T cells. Maturing thymocytes lose either CD4 or CD8 surface molecules and are exported to peripheral tissue as CD4+ T_H cells or CD8+ CTLs.

with B-cell development, the developmental stages of T cells correspond to gene rearrangement events, culminating in specificity for the TCRs. The TCR gene rearrangement is referred to as **V(D)J recombination**. As T cells develop, germ-line sequences undergo recombination events of specific genes in the V (variable), D (diversity), and/or J (joining) regions (Fig. 4.3). Similar to B cells, N region diversification can occur whereby extra nucleotides are inserted by the enzyme deoxynucleotidyl transferase; this expands the potential pool of TCRs generated. The rate of T-cell production in the thymus is highest in young individuals. The thymus shrinks during adulthood, suggesting that the complete T-cell repertoire is primarily established prior to puberty.

The thymus plays an integral role in both education and regulation of T-cell lineage, with development starting in the thymic cortex and ending in the medulla. Progenitor cells migrating into the thymus do not express the dominant **CD4** and **CD8** markers. Thus, these progenitor cells are termed "double-negative" thymocytes. The development step is rearrangement of the TCR, with commitment to the αβ T-cell or γδ T-cell

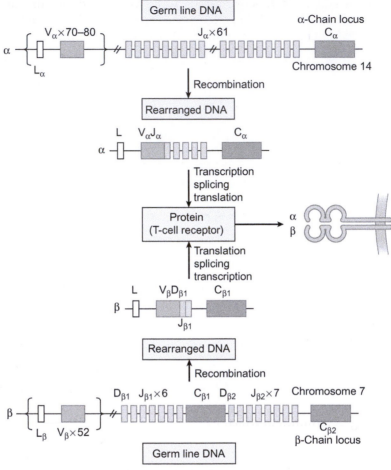

FIG. 4.3 Gene rearrangement of TCR loci. TCR diversity is generated by the combinatorial joining of variable (V), joining (J), and diversity (D) genes and by N-region diversification (nucleotides inserted by the deoxynucleotidyl transferase enzyme). The top and bottom rows show germ-line arrangement of the V, D, J, and constant (C) gene segments at the TCRα and TCRβ loci. During T-cell development, variable sequences for each chain are assembled by DNA recombination. For the α chain *(top)*, a Vα-gene segment rearranges to a Jα-gene segment to create a functional gene encoding the V domain. For the β chain *(bottom)*, the rearrangement of Dβ-, Jβ-, and Vβ gene segments creates the functional V domain exon.

phenotype. As previously described for B lymphocytes, T lymphocytes are also created with a vast array of antigenic specificities prior to contact with antigen. During TCR gene rearrangement, genes undergo recombination. The developing cells must express a functional TCR to survive. At this stage, the CD3 molecule is also expressed, allowing intracellular signal transduction. This signaling causes proliferation and expression of CD4 and CD8. Cells are now referred to as "double-positive" thymocytes.

Once the full αβ TCR is expressed on the thymocyte cell surface, the selection process that will shape the T-cell repertoire begins. T-cell recognition of antigen requires recognition of not only the peptide epitope presented, but also the antigen presentation molecule itself, the **major histocompatability complex (MHC)**. The first selection process is positive selection, and the new αβ TCR that recognizes MHC and the peptide epitope is allowed to mature. The positive recognition of MHC class I or class II coordinates with development of CD4 and CD8 T cells, respectively. This means that a thymocyte that recognizes MHC class I will cease expression of CD4, leading to the development of a CD8 T cell, and vice versa. Thus, at the end of positive selection, the αβ TCR expression increases, and the cell becomes a single positive thymocyte. The next selection process, negative selection, where the cells are tested for reactivity against self-antigens. Those that survive finish the maturation process and are released into circulation.

ANTIGEN RECOGNITION BY T CELLS: REQUIREMENT OF MAJOR HISTOCOMPATABILITY MOLECULES

T cells only recognize antigens that are processed into short peptides and presented on the surface in an antigen presentation molecule, the MHC, which is referred to as **human leukocyte antigen (HLA)** in humans. **MHC class I** molecules are found on all nucleated cells, whereas **MHC class II** are only present on professional **antigen-presenting cells [APCs;** macrophages, dendritic cells (DCs), and B cells]. The method of peptide processing determines to which MHC it will be loaded.

THE HLA LOCUS

The HLA locus in humans is found on the short arm of chromosome 6 (Fig. 4.4). The class I region consists of HLA-A, HLA-B, and HLA-C loci. The class II region consists of the D region, subdivided into HLA-DP, HLA-DQ, and HLA-DR subregions. A region between the class I and class II loci encodes for class III proteins with no structural similarity to either class I or class II molecules. The class III molecules include complement proteins, tumor necrosis factor, and lymphotoxin.

The highly polymorphic class I and class II MHC products are central to the ability of T cells to recognize foreign antigen and the ability to discriminate the "self" from the "nonself." MHC class I and class II molecules that are not possessed by an individual are seen as foreign antigens upon transplantation and are dealt with accordingly by the recipient's immune system (discussed in Chapter 11). All MHC molecules show a high level of

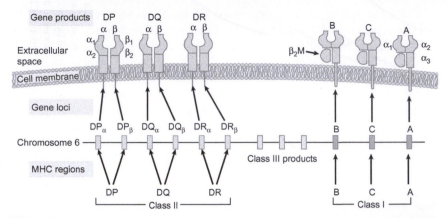

FIG. 4.4 Genetic organization of the HLA locus and associated gene products. The polymorphic human MHC genes of the HLA locus coding for class I and class II molecules are located on chromosome 6. Class I genes are designated as A, B, and C, each coding for a three-domain polypeptide (α1, α2, and α3) associated with invariant β_2-microglobulin (β_2M). The class II genes are DP, DQ, and DR, each coding for individual α and β chains that interact to provide binding sites for antigen presentation.

allotypic polymorphism (i.e., certain regions of the molecules differ from one person to another). The chance of two unrelated people having the same allotypes at all genes that encode MHC molecules is very small.

The MHC class I molecules are each somewhat different from one another with respect to amino acid sequence, and all three (HLA-A, HLA-B, and HLA-C) are codominantly expressed on all nucleated cells. **Codominantly expressed** means that each gene encoding these proteins from both parental chromosomes is expressed. The MHC class II molecules are a bit different, in that expression includes homologous and heterologous $\alpha\beta$ dimer mixtures, representing proteins from both parents. Both α and β subunit genes exhibit species-specific polymorphism. Homologous dimers match class II molecules expressed on either parental cell type, whereas heterologous dimers are unique to the F1 genotype and are functionally nonequivalent to parental class II molecules.

MHC CLASS I

MHC class I molecules bind **endogenous**-derived peptides from antigens processed in the cytosol (Fig. 4.5), such as viral proteins produced within the host cell. Protein degradation occurs naturally by a complex called the "proteasome." Antigens are degraded and transported to the endoplasmic reticulum (ER) by the **transporters associated with antigen processing (TAP)** protein, and the peptides are further processed and

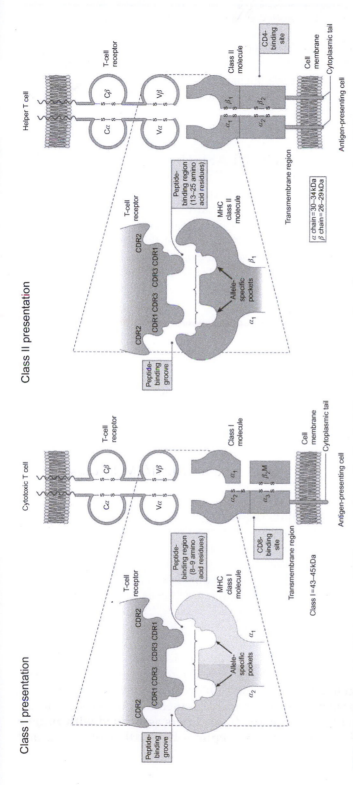

FIG. 4.5 Endogenous and exogenous MHC presentation to T cells. (*Left*) The endogenous pathway for antigen processing allows MHC class I molecules to interact with intracellular-produced peptides degraded by proteasomes. The MHC class I transmembrane molecule is associated with the invariant β₂-microglobulin β₂M, giving structure to the extracellular domain for the presentation of processed antigens to CTLs. Short peptide fragments (8–10 amino acids in length) noncovalently interact with the domains on the class I molecule and complementary determining regions (CDRs) on the TCR, stabilized by the CD8 molecule. (*Right*) The exogenous pathway prepares processed antigens for presentation to CD4+ T cells via MHC class II-regulated mechanisms. Processed antigenic fragments (13–25 amino acids in length) interact with domains on the class II molecule within the peptide-binding groove, allowing presentation to CD4+ T-helper cells. Interactions with the TCR are stabilized by CD4 recognition of conserved regions on the class II molecule.

loaded onto the MHC class I. Once the peptide is bound to MHC class I, the complex is stabilized by a **β2-macroglobulin** molecule and then exported to the cell membrane surface. The entire surface molecule can now be recognized by CD8 T cells (cytotoxic T cells) that are specific for the bound peptide.

MHC CLASS II

MHC class II molecules bind **exogenous**-derived peptides from antigens processed in organelles (Fig. 4.5). Any extracellular antigen, such as whole bacteria, engulfed by APCs via phagocytosis or endocytosis is enclosed in an intracellular vesicle. Activation of the APCs leads to acidification and activation of proteases to degrade the antigen into peptide fragments. The vesicle containing peptides is fused with another vesicle containing MHC class II proteins. Upon fusion, the peptide is loaded on to MHC class II molecules, and the entire complex migrates to the cell membrane surface, where peptide specific CD4 T cells (helper T cells) can recognize it.

T LYMPHOCYTE FUNCTIONS

To generate an active immune response, a small number of B- and T-cell clones that bind antigen with high affinity undergo activation, proliferation, and differentiation. Some of these cells become **effector T cells** that express activities that help to eliminate the pathogen. Others become **memory cells** that can give rise to secondary responses as described next.

Complete cellular activation requires a series of interactions (Fig. 4.6). Naive T cells that migrate out of the thymus circulate through the peripheral lymphoid organs, searching for cognate antigen recognized via specific surface molecules. These naive T cells carry surface receptors to allow targeted migration to lymph nodes, where they make contact with resident DCs. The DCs represent a population of cells that specifically promote recognition of small linear peptide epitopes, functioning as APCs to promote further development of naive T cells. It is of note that APCs may be found in various tissues throughout the body, often with different names dependent on location where they reside. For example, **Langerhans cells** are dendritic cells of the skin, **microglia** reside in the brain, **Kupffer cells** line the sinusoids of the liver, and **dust cells** are located in the pulmonary alveoli.

Activation requires three signals. The first is binding of the TCR to the peptide/MHC. The second is binding of costimulatory molecules

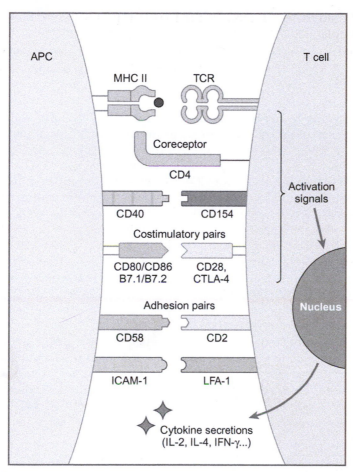

FIG. 4.6 Interactions between the T cell and APC. Activation of lymphocytes requires multiple signals that include binding of the antigen-specific receptor to the antigen, interaction with stabilizing surface molecules, and exposure to cytokines or secondary costimulatory signals. For the T-helper cells shown, recognition of the presented antigen is enhanced by CD4, which attaches to nonpolymorphic regions on the MHC class II molecule. Further stabilization is accomplished by integrin interactions between molecules on the T cell with ligands on the surface of the APC. Costimulatory molecules present on both the T cell and the APC are critical to T-cell function and activation.

(CD28 on T cells to CD86 or CD80 on DCs). The third signal includes cytokines produced by the DCs during encounter with the naive T cell. It is this third signal that is responsible for T-cell differentiation. Once activated, the effector T cells proliferate, develop specific subtype features, and express receptors to allow them to migrate to sites where activity is needed.

CD4+ T-HELPER CELLS

CD4 T cells recognize antigenic epitopes presented by MHC class II molecules. Upon activation by APCs, the CD4 T cells proliferate and differentiate into several subtypes, each with specific functional capabilities (Fig. 4.7).

T-helper type 1 (TH1) cells are effector CD4 T cells that differentiate when interleukin-12 (IL-12) is given as the third signal. T_H1 cells produce primarily IFN-γ cytokine, which is a primary cytokine used to activate macrophages to promote intracellular killing of pathogens. The **T-helper type 2 (TH2)** cells are effector CD4 T cells that differentiated in the presence of interleukin-4 (IL-4) and produce vast quantities of IL-4, interleukin-5 (IL-5), and interleukin-13 (IL-13). T_H2 cells provide help to promote antibody production, specifically IgE, to target helminthes and parasites. Indeed, the original name for IL-4 was B-cell growth factor. It should be noted that other major subsets of T cells exist; formation of these subsets is also under cytokine regulation. For example, **T-helper 17 (TH17)** cells are differentiated under TGF-β1 and interleukin-6 (IL-6); they produce high levels of interleukin-17 (IL-17) and enhance neutrophilic responses.

One major function of CD4 T cells is to help activate B cells and promote antibody production. It is now hypothesized that the main effector CD4 cell responsible for this process in the lymphoid follicles is the **T follicular helper cell (TFH)**. The T_{FH} cell differentiates under the presence of IL-6 and can secrete a wide variety of cytokines, thus enabling

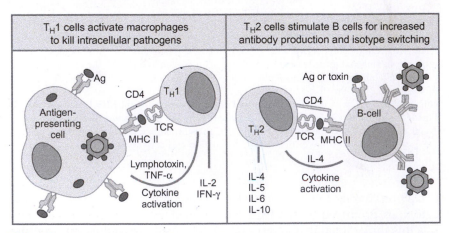

FIG. 4.7 Effector functions of T-helper cells. T_H1 cells activate macrophages through the production of cytokines, leading to assisted destruction of intracellular microorganisms. T_H2 cells drive B-cell differentiation to stimulate B cells to proliferate, to secrete immunoglobulins, and to undergo isotype switching events. *IFN*, interferon; *IL*, interleukin.

stimulation of production of antibodies associated with both T_H1 (IgG2a) and T_H2 (IgE, IgG1) immune responses.

Regulatory CD4 T cells (**Tregs**) are a unique subset of lymphocytes. T cells produce FoxP3, a transcription factor, and phenotypically express CD25 molecules on their cell surface. These regulatory T cells affect effector functions by several methods that include direct cell-to-cell contact, increased expression of surface molecules, increased responsiveness to growth factors, and altered production of regulatory cytokines such as TGF-β1 and interleukin-10 (IL-10). Finally, other phenotypic populations of T-helper cells exist (**TH9, TH22**), which also control unique facets of immune response.

EVENTS INVOLVED IN T LYMPHOCYTE ACTIVATION

Immune response must be regulated to allow sufficient activity to protect the host without excessive or inappropriate responses (hypersensitivities, which are discussed further in Chapter 8) that may lead to disease states. Receptor recognition of antigen mediates the transcription of cytokine genes by a complex sequence of molecular events. Absence of a secondary signal can lead to cellular inactivation.

The professional presenting cells that provide costimulation are DCs, macrophages, and B cells. Of these cell types, DCs deliver the best costimulatory signals for activation of naive T cells. Presenting cells use the membrane molecules of the B7 family to deliver costimulatory signals through interaction with their ligand on T-cell membranes, CD28; two isoforms of the B7 family (B7.1 and B7.2, which are also called CD80 and CD86) act to regulate cytokine responses. For example, CD80 binding upregulates T_H1 cytokines [interleukin-2 (IL-2), IFN-γ], while CD86 binding upregulates T_H2 response (IL-4, IL-5, IL-6, and IL-10). Interactions with CD28 are critical, in that binding to CD28 on the lymphocyte leads to IL-2 production, a major T-cell growth factor, while inhibition of binding leads to development of tolerance. Reactivity leads to differential intracellular regulation of transcription factors. Interaction of the B7 molecules with other T-cell surface antigens, such as CD152 (CTLA-4), leads to suppressive signals and tolerance/anergy, and the induction of memory cell formation (Fig. 4.8).

Macrophages, as presenting cells, help prime the environment to prepare the incoming T cell for phenotypic differentiation. The "classical" M1 type of macrophage produces proinflammatory mediators to augment T_H1 cells to protect against bacteria and viruses, as well as intracellular threats. The M2 macrophages represent a broader phenotypic subset that is more closely involved with T_H2 differentiation for larger parasite control and antibody production. The entire process requires a coordinated

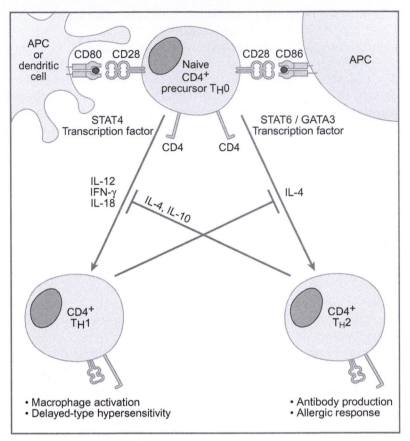

FIG. 4.8 Development of T-helper-cell phenotypes. T_H1 and T_H2 cells derive from precursor cells under the influence of local cytokines. Each secretes a phenotypic subset of cytokines that drive effector responses. Secreted cytokines modulate response to the other subset and inhibit alternative functional development. Not shown are the T_H17 and T regulatory subsets, which mature under the influence of cytokine IL-17 and TGF-β.

response between cells that is regulated by signaling through recognition receptors in the presence of unique cytokine subsets.

Realize that it is critical to have built-in signals for regulating immune activation. **Immune checkpoints** are molecules designed to regulate **immune** activation and play key roles in maintaining homeostasis and preventing reactivity to self (autoimmunity). Activated T cells express surface molecules that are involved in inhibition in a feedback manner. One of these, CTLA-4, was mentioned earlier in this discussion. Two others include the lymphocyte-activation gene 3 (LAG-3) and the programmed cell death protein 1 (PD-1). These checkpoint molecules also function to enhance antitumor capabilities.

ROLE OF T CELLS IN B-CELL ACTIVATION

B cells are capable of using their surface immunoglobulin to engulf anti-gen, after which they can process and present antigenic fragments to T cells. This promotes the direct elicitation of cytokines from T cells for stimulation, resulting in activation and plasma cell development (Fig. 4.9). Clonal expansion of B cells under influence of T-cell cytokines leads to plasma-cell development, isotype class-switching, and produc-tion of memory responses. Specifically, IL-4 secreted from T$_H$2 cells acts as a B-cell growth factor, and IL-6 assists in delivering signals for matura-tion of the antibody response. It is important to keep in mind that the inter-action between T cells and B cells is complex; B cells secrete a large number of proinflammatory mediators, including IL-2, TNF-α (tumor necrosis fac-tor alpha), and IL-12, which have the potential to modulate T-cell prolif-eration and phenotypic function. However, under certain circumstances, B cells can also respond and proliferate through **T-independent** mecha-nisms, usually involving antigens with long-repeating epitopes that allow the cross-linking of immunoglobulin receptors on the surface of the B cell.

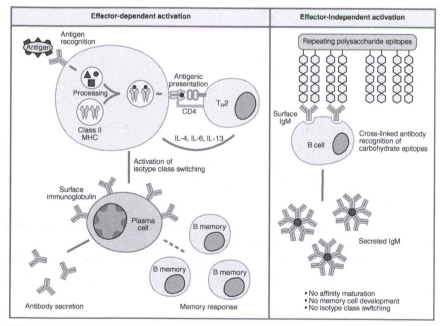

FIG. 4.9   B cell response to T-dependent and T-independent antigens. *(Left)* T$_H$2 cells spe-cifically recognize class II-presented antigenic determinants and drive B-cell activation, lead-ing to isotype class-switching and antibody secretion. *(Right)* Alternatively, T-independent responses to repeating carbohydrate epitopes stimulate antibody production but do not lead to maturation of the antibody response.

A common example occurs with bacterial capsid antigens containing repeating carbohydrate (polysaccharide) epitopes. In the case of T-independent activation, there is no accompanying maturation of response; antibody production is primarily limited to IgM isotypes.

CYTOTOXIC T-CELL EFFECTORS

CD8+ T Cells

CD8 T cells recognize antigens presented on MHC class I molecules and become **cytotoxic CD8 T cells (CTLs)** when activated. While CD8 T cells also produce various cytokines, their main activity is to eliminate infected host cells. Activation of CD8 T cells requires additional costimulatory signals. This can occur with or without the help of CD4 T cells. In some instances with infected DCs, the infection creates enough inflammation for their production of cytokines and costimulatory signals that are sufficient to activate CD8 T cells in the absence of CD4 T cells. However, most CD8 T-cell activation requires CD4 T cells to sufficiently help activate and upregulate costimulatory signals that are required for optimized target cell destruction.

Activated CD8 T cells kill target cells by apoptosis (Fig. 4.10). One major method is through the release of cytotoxic granules. Binding of the TCR to

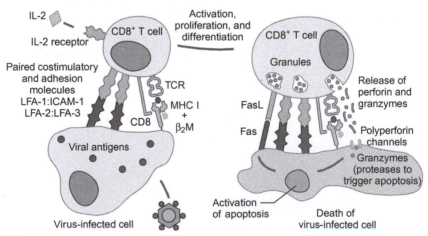

FIG. 4.10 Effector function of CTLs. CTLs recognize virally infected target cells that express foreign antigens complexed with MHC class I molecules. Responses are mediated through IL-2 and the IL-2 receptor (CD25), and strengthened through interactions between the CTL and the infected target cell. Cytotoxic effector molecules produced (i.e., perforins, granzymes) initiate the destruction of the target cell and deliver apoptotic signals through Fas and FasL on cellular surfaces. *LFA*, lymphocyte function-associated antigen; *ICAM*, intracellular adhesion molecule.

MHC class I triggers the synthesis of **perforin** and **granzymes**, which are stored within the cytosol. The granules are released at the point of contact, allowing specific targeting and limited bystander death. Perforin assembles on target membranes, allowing the delivery of granzymes into the target cell. Granzymes are a group of serine proteases that activate caspases, leading to cell death. Direct cell-to-cell contact is also critical for functionality. CD8 cells can induce apoptosis by ligation of **Fas** and **Fas ligand**, which are expressed on lymphocytes and on infected target cells. Activated CD8 T cells also produce several cytokines that contribute to host defense, including IFN-γ, TNF-α, and lymphotoxin-α. IFN-γ inhibits viral replication while increasing expression of MHC class I, improving the chance that an infected cell will be recognized.

While the majority of CD8 T cells function to control the growth of intracellular pathogens (e.g., virus-infected cells), they also play a major role in natural surveillance and in limiting the growth of cancerous cells. Finally, subpopulations of CD8 cells with regulatory/suppressive function have been identified recently, and perhaps they are involved in self-tolerance. These CD8 cells are different in phenotype and function from the CD4 Treg cells.

INNATE LYMPHOCYTES AND SUPERANTIGENS

Up to now, the lymphocytes described have all been involved in the adaptive immune response. This means that upon the introduction of a new antigen, such as during the initial infection or vaccination, the lymphocytic response requires approximately 2 weeks to fully activate the appropriate naive cell, undergo clonal expansion, and initiate cellular migration to the site of infection to enact activity. However, there are lymphocytes that operate in the innate capacity and appear at the site of infection or inflammation within 1–2 days of infection. These innate lymphocytes make up a minor population of lymphocytes that recognize only a small number of unique antigens.

γδ T CELLS

The antigen receptors of some T cells are comprised of γ and δ polypeptides rather than the common α and β chain TCRs described previously. These γδ T cells constitute 10%–15% of the human blood T-cell population and are abundant in the gut epithelia, the lungs, and the skin. They appear to be important to immune responses to epithelial pathogens. Unlike conventional αβ T cells, γδ T cells do not go through positive or negative selection and are released into circulation upon successful rearrangement of their TCRs. Generally, γδ cells lack CD4, although some express CD8. Unlike αβ T cells, γδ T cells generally do not recognize antigens presented on MHC molecules, but they do recognize antigenic targets directly, similar to antibodies. Their function is not well known, but it is hypothesized that they recognize changes in heat-shock proteins, MHC class Ib molecules, and phospholipids in infected cells. In this way, they may suffice to mount rapid responses against a subset of phylogenetically conserved ligands, thus representing an efficient early defense against pathogenic organisms.

NATURAL KILLER T (NKT) CELLS

Natural killer T (NKT) cells are a heterogeneous group of T cells that share properties of both T cells and natural killer (NK) cells. The majority of these cells recognize an antigen-presenting molecule (CD1) that binds self and foreign lipids and glycolipids (Fig. 4.11). CD1 molecules are non-

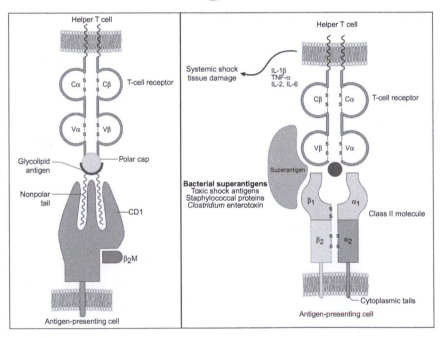

FIG. 4.11   *(Left)* Nonclassical lipid antigen presentation by CD1 molecules. The peptide groove of the CD1 surface molecule is lined with nonpolar, hydrophobic side chains that allow the binding of antigen in a deep, narrow hydrophobic pocket to enable presentation to the TCR. *(Right)* Superantigens. T cells of various antigenic specificities are activated when bacterial superantigens cross-link MHC class II molecules with common TCR Vβ regions. This causes activation in the absence of specific peptide filling the MHC molecule.

MHC restricted and nonpolymorphic. However, although distinct from MHC classes I and II, they show a similar structure to MHC class I, having three extracellular domains and expressed in association with β2-microglobulin. They constitute only 0.2% of all peripheral blood T cells. It is now generally accepted that the term "NKT cells" refers primarily to CD1-restricted T cells coexpressing a heavily biased, semiinvariant TCRα chain paired with one of three TCRβ chains. Upon activation, these cells secrete several cytokines, including IFN-γ, IL-4, and IL-10, and perform direct killing of target cells. NKT cells should not be confused with NK cells.

SUPERANTIGENS

A "superantigen" is a molecule that is able to elicit T-lymphocyte responses by circumventing normal antigen processing and presentation functions. Superantigens are defined by their ability to stimulate a large fraction of T cells via interactions with the TCR Vβ domain (Fig. 4.11). Superantigens are predominantly bacterial in origin, such as staphylococcal enterotoxins and toxin-1, which are responsible for toxic shock syndrome. Superantigens directly bridge the TCR with the MHC class II molecule, causing T cells to divide and differentiate into effector cells to release cytokines [IL-2, TNF-α, and interleukin-1 beta (IL-1β)]. Because the number of T cells that share Vβ domains is high (up to 10% of all T cells), large numbers of cells may be activated, regardless of antigen specificity. This leads to massive systemic disruption and clinical features similar to septic shock.

SUMMARY

- T lymphocytes are regulators of adaptive function, serving as primary effectors for cell-mediated immunity. Antigenic specificity is dictated by means of the TCR heterodimer receptor, derived from the recombination of gene segments.
- Recognition of presented antigens in the context of major histocompatability molecules and costimulatory molecules dictates the differentiation of effector processes.
- CD4+ T-helper lymphocyte cells recognize exogenous antigens presented in the context of MHC class II molecules. Different subclasses of T-helper cells secrete unique subsets of cytokines that assist in functional activity.

- CD8+ T lymphocytes, also known as CTLs, recognize endogenous antigens presented in the context of MHC class I molecules. CTLs kill target cells directly by inducing apoptosis via released, preformed proteins.
- Other mechanisms for T-cell activation have evolved, most likely in response to pathogens that circumvent typical pathways required for adaptive immune protection.

How We Defend Against Infectious Agents

CHAPTER FOCUS

To examine the process of defense against different types of pathogens, with an element focused on the commonly encountered organism classes. Opportunistic infections are discussed, especially about how they are related to the diminished response induced when individuals are immuno-compromised. General mechanisms that organisms use to evade immune function will also be addressed.

IMMUNE HOMEOSTASIS AND PATHOGENIC ORGANISMS

The human host represents an immense microbiome, with hundreds of trillions of symbiotic microorganisms living on or in our bodies. Yet our bodies are exquisitely adapted to handle constant assault by various classes of pathogens and opportunistic agents. Indeed, we are able to balance normal commensal flora with the ability to distinguish specific agents that are harmful to daily living, a feature that antibiotics do not share. Indeed, the immune system is able to balance responses to its normal microbiome with the need to control and eliminate infectious agents that cause disease. This is done at the innate level of response, with activity depending on physical barriers and cellular recognition of pathogenic motifs. It is also accomplished at the adaptive level, with reactivity directed toward specific foreign antigens on pathogenic invaders.

MAJOR IMMUNE DEFENSE MECHANISMS AGAINST PATHOGENS

The course of response against typical acute infections can be subdivided into distinct stages. Initially, the level of infectious agent is low, beginning with breach of a mechanical barrier (e.g., the skin or mucosal surfaces). Once inside the host, the pathogen encounters a microenvironment for suitable replication. The agent replicates, releasing antigens that trigger innate immune function, generally characterized as nonspecific. Preformed effector molecules recognize microorganisms within the first 4h of infection and assist in limiting the expansion of the organism. Complement components and released chemokines attract professional phagocytes [macrophages and polymorphonuclear cells (PMNs)] and natural killer (NK) cells to the site of infection in order to assist in activation of these cells. After 4 or 5days, antigen-specific lymphocytes (B and T cells) undergo clonal expansion, enabling directed control and eventual clearance of the infectious agent. The host is left with residual effector cells and antibodies, as well as immunological memory to provide lasting protection against reinfection.

A wide variety of pathogenic microorganisms exist. They may be globally classified into groups: bacterial, mycobacterial, viral, protozoal, parasitic worms, and fungal. The host defense is based on the availability of resources to combat a localized pathogen (Fig. 5.1).

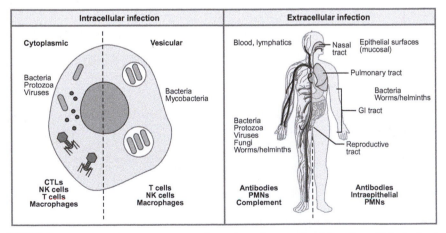

FIG. 5.1   Functional immune response depends on the organism's location within the host. Effective immune responses are directed against intracellular organisms residing in cytoplasmic or vesicular space, or against extracellular organisms residing at mucosal surfaces or present in blood, lymph, or tissue.

PHYSICAL BARRIERS TO INFECTION

Four major categories of physical barriers exist to limit entry and control expansion of foreign pathogens. Defensive roles may be **anatomic** (skin, mucous membranes), **physiologic** (temperature, low pH, chemical mediators), **phagocytic** (digestion of microorganisms), or **inflammatory** (vascular fluid leakage).

Anatomic Barrier

The skin has the thin outer epidermis and the thicker underlying dermis to impede entry and provide an effective barrier against microorganisms. Sebum, produced in sebaceous glands, is made of lactic acid and fatty acids that effectively reduce skin pH to between 3 and 5, which inhibits organism growth. Mucous membranes are covered by cilia that trap organisms and propel them out of the body.

Physiologic Barrier

The physiologic barrier includes factors such as temperature, low pH, and chemical mediators. Many organisms cannot survive or multiply in elevated body temperature. Soluble proteins play a major role in innate responses. Lysozymes can interact with bacterial cell walls; interferons alpha (α) and beta (β) are natural inhibitors of viral growth; complement components use both specific and nonspecific immune factors to convert inactive forms to active moieties that damage the membranes of pathogens. Low pH in the stomach and gastric environment discourages bacterial growth.

Phagocytic and Endocytic Barriers

Blood monocytes, tissue macrophages, and neutrophils phagocytose and kill microorganisms via multiple complex digestion mechanisms. Bacteria become attached to cell membranes and are ingested into phagocytic vesicles. Phagosomes fuse with lysosomes where lysosomal enzymes digest captured organisms.

Inflammatory Barriers

Initial localized tissue damage by invading organisms results in the release of chemotactic factors (complement components, fibrinopeptides) to signal changes in nearby vasculature that allow the **extravasation** of

PMNs to the injury site. Innate immune recognition utilizes preformed effector molecules to recognize broad structural motifs that are highly conserved within microbial species. Engagement of innate components leads to the triggering of signal pathways to promote inflammation, ensuring that invading pathogens remain in check while the specific immune response becomes engaged.

Virtually all pathogens have an extracellular phase where they are vulnerable to antibody-mediated effector mechanisms. If an extracellular agent resides on epithelial cell surfaces, antibodies such as immunoglobulin A (IgA) and nonspecific inflammatory cells may be sufficient for combating infection. If the agent resides within interstitial spaces, in blood or in lymph, then protection also may include macrophage phagocytosis and neutralization responses. Intracellular agents require a different response to be effective. For cytoplasmic agents, T lymphocytes and NK cells, as well as T cell-dependent macrophage activation, are usually necessary to kill the organisms.

Pathogens can damage host tissue by direct and indirect mechanisms. Organisms may directly damage tissue by releasing exotoxins that act on the surface of host cells or endotoxins that trigger local production of damaging cytokines. Pathogens also may directly destroy the cells that they infect or force indirect damage through the adaptive immune response. Pathological damage may occur due to excess deposition of antibody:antigen complexes or through bystander killing effects during overactive specific responses toward infected host target cells (as discussed in Chapter 8).

BACTERIAL INFECTIONS

Bacterial infections begin with a breach of a mechanical barrier, after which released bacterial factors during replication initiate a cascade of events (Fig. 5.2). Initially, infection may be resisted by antibody-mediated immune mechanisms, including neutralization of bacterial toxins. However, the role of complement in response to bacterial infection must be stressed. Major biological components of the complement system include activation of phagocytes, direct cytolysis of target cells, and coating (opsonization) of microorganisms for uptake by cells expressing complement receptors.

Complement is a system of more than 30 serum and cell surface proteins involved in inflammation and immunity. In conjunction with specific antibodies, complement components act as a primary defense against bacterial (and viral) infections. Most of the complement proteins are "acute-phase proteins" produced by liver hepatocytes and found in the serum. Complement components can increase in concentration

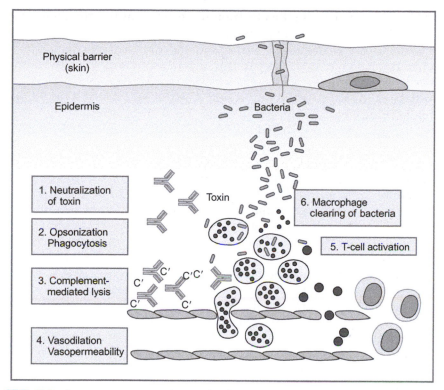

FIG. 5.2 Immune response to bacterial infections. Immune defenses against bacterial agents include antibodies for the neutralization of toxins, opsonization of organisms for targeted destruction, and activation of complement for direct lysis. Vasodilation of blood vessels allows the entry of cells to sites of infection to assist in control of infection.

twofold-to-threefold during infection. The sequential activation of complement proteins, called the **complement cascade**, is initiated by protein-protein interactions. At each step, the number of activated molecules increases, amplifying the reaction. Many complement proteins are present as zymogens (inactive precursors), which are activated by either conformational changes or proteolytic cleavage by other complement proteins. Activation of these zymogens results in specific serine protease activity capable of cleaving other complement proteins, producing the complement cascade (see Figs. 2.2 and 2.3).

Several complement polypeptide cleavage fragments are involved in primary biological functions of immunity. Their main function begins when antibodies recognize pathogenic determinants and form the basis of a physical structure to which complement components interact. Specifically, complement components interact with the Fc portion of immunoglobulin M (IgM) and immunoglobulin G (IgG) binding to the surface of bacteria. This initiates a cascade of events whereby a membrane attack complex (MAC) is built upon the cellular surface, assembling as a pore channel in the lipid bilayer membrane and causing osmotic cell lysis. MAC formation requires prior activation by either the classical or alternative pathway and utilizes the proteins C5b, C6, C7, C8, and C9.

Related complement components [C3b or inactivated C3b (iC3b)] can bind directly to pathogens. Interaction with complement receptors on the surface of macrophages, monocytes, and neutrophils leads to enhanced phagocytosis and targeted destruction of organisms. Proteolytic degradation of C3 and C5 leads to the production of leukocyte chemotactic factors referred to as **anaphylatoxins**. For example, C3a is chemotactic for eosinophils. C5a is a much more potent chemokine, attracting neutrophils, monocytes, macrophages, and eosinophils. Interaction of C3a, C4a, or C5a with mast cells and basophils leads to the release of histamine, serotonin, and other vasoactive amines, resulting in increased vascular permeability, causing inflammation and smooth muscle contraction. The vasodilatation and vasopermeability result in an influx of professional phagocytes and acute polymorphonuclear infiltrates.

Neutrophils are typically the first infiltrating cell type to the site of inflammation (Fig. 5.3). Activated endothelial cells increase the expression of E-selectin and P-selectin, which are recognized by neutrophil surface mucins (PSGL-1 or sialyl Lewis^x on glycoproteins or glycolipids) and induce neutrophil rolling along the endothelium. **Chemoattractants** such as CXCL8 and interleukin-8 (IL-8) can further trigger firm adhesion and diapedesis. Subsequent chemotaxis can also be induced by fibrinopeptides and leukotrienes. Activated neutrophils express high-affinity Fc receptors and complement receptors that allow increased phagocytosis of invading organisms. The activation of neutrophils leads to a respiratory burst that produces reactive oxygen and nitrogen intermediates, as well as releasing primary and secondary granules containing proteases, phospholipases, elastases, and collagenases. Pus, a yellowish white opaque creamy matter produced by the process of suppuration, consists of innumerable neutrophils and tissue debris.

Macrophages involved in innate immunity include **alveolar macrophages** in the lung, **histocytes** in connective tissue, **Kupffer cells** in liver, **mesangial cells** in kidney, **microglial cells** in brain, and **osteoclasts** in bone.

Chemokine mediators such as macrophage inflammatory protein-1α (MIP-1α) and macrophage inflammatory protein-1β (MIP-1β) attract monocytes to the site of pathogenic infection. Like the neutrophil, monocytes express surface ligands called "integrins," which recognize ligands (cell adhesion molecules) on endothelial cells. This interaction mediates cell rolling, firm adhesion, and **diapedesis**. The entire process is referred to as **extravasation**. Activated tissue macrophages secrete proinflammatory

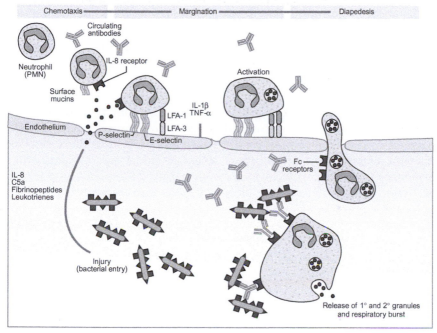

FIG. 5.3 Events associated with neutrophil transendothelial migration. Bacteria entering through a breach in the mechanical barrier (skin) trigger the release of chemotactic factors that upregulate selectins on endothelial beds. Circulating PMNs interact via weak-binding surface mucins; interactions are enhanced in the presence of stimulating cytokines, leading to margination and diapedesis. The neutrophils shown enter tissue and undergo a respiratory burst to release primary and secondary granules upon cross-linking of Fc receptors with antibodies recognizing bacterial epitopes.

mediators in response to bacteria and bacterial products, including interleukin-1 (IL-1), interleukin-6 (IL-6), IL-8, interleukin-12 (IL-12), and tumor necrosis factor alpha (TNF-α). TNF-α is an inducer of a local inflammatory response to contain infections. IL-1, IL-6, and TNF-α play a critical role in inducing the acute phase response in the liver and fever, which favor effective host defense in several ways. In addition, IL-12 may activate NK cells.

Microbial Motifs Detected Through Pattern-Recognition Receptors

Monocytes and macrophages express receptors that recognize broad structural motifs called **pathogen-associated molecular patterns (PAMPs)**, which are highly conserved within microbial species. Such receptors are also referred to as **pattern-recognition receptors (PRRs)**.

Engagement of these receptors leads to the immediate triggering of signal pathways that promote phagocytosis, an event that requires actin and myosin polymerization. Multiple factors assist in preparing the particulate for engulfment and targeting for destruction, including various opsonins comprised of complement components. Examples of PRRs include receptors that recognize common bacterial carbohydrate elements, such as lipopolysaccharide (LPS), mannose, and glucans (glycogen polysaccharides). A unique family of membrane-bound receptors, termed **Toll-like receptors** (**TLRs**), plays a critical role in the recognition of bacterial components (cell-wall membranes, LPS, flagellin, CpG oligodeoxynucleotides); interaction with any of these receptors initiates proinflammatory responses that function locally and systemically (Table 5.1). In the case of overwhelming infection, this class of PRRs is actively engaged, contributing to the presentation of sepsis, clinically manifested as an overwhelming proinflammatory cascade. Of note, additional diversity occurs when different TLRs dimerize; the heterodimers function to increase the recognized motifs. Finally, it is important to note that while many TLRs are located on the cell surface, a number of them are located intracellularly. This gives the advantage of triggering immune responses to released RNA or DNA that is accessible only after the infectious agent has broken down inside the cell.

The process of phagocytosis compartmentalizes the invading pathogen into an intracellular vacuole referred to as a **phagosome**. The phagosome fuses with intracellular lysosomes, forming a **phagolysosome** vacuole. The lysosome is extremely acidic in nature and contains powerful lytic enzymes; fusion with the phagosome allows directed delivery for targeted destruction of invading organisms. In addition, phagocytosis triggers a mechanism known as the **respiratory burst**, leading to the production

TABLE 5.1   TLRs and Their Ligands

TLR	Location	Ligand
TLR1, TLR2, TLR6	Extracellular	Lipopeptides
TLR3	Intracellular	Double-stranded RNA
TLR4	Extracellular	Gram-negative LPS
TLR5	Extracellular	Flagellin
TLR7, TLR8	Intracellular	Single-strand RNA
TLR9	Intracellular	Unmethylated CpG DNA

Note: Listed here are the important TLRs and specific ligands involved in pathogen recognition, as well as their cellular localization. At least 15 different TLRs have been identified, with ligand motifs identified for most of them.

of toxic metabolites that assist in the killing process. The most critical of these toxic metabolites are nitric oxide, hydrogen peroxide, and superoxide anions.

During the chronic stage of the infection, cell-mediated immunity (CMI) is activated. T cells reacting with bacterial antigens may infiltrate the site of infection, become activated, and release lymphokines that further attract and activate macrophages. Likewise, NK cells enter the infected region and assist in macrophage activation. The activated macrophages phagocytose and degrade necrotic bacteria and tissue, preparing the lesion for healing. The PMNs, especially neutrophils, are an excellent example of the first line of innate defense against bacterial agents.

Important factors released by macrophages in response to bacterial antigens include cytokines that exert both local and systemic function. A major outcome of triggering through recognition of pathogenic motifs is the production of inflammatory cytokines, chemokines, and antimicrobial peptides. Locally, IL-1, TNF-α, and IL-8 cause inflammation and activate vascular endothelial cells to increase permeability and allow more immune cells to enter infected area. TNF-α will also destroy local tissue to limit the growth of bacteria. In addition, IL-6 can stimulate an increase in B-cell maturation and antibody production, and IL-12 will lead to the activation of NK cells and priming of T cells toward a T_H1 response. The stimulation also triggers the expression of a special class of molecules called **interferons**—specifically, Type I interferon-alpha and Type I interferon-beta (IFN-α and IFN-β). Systemically, interleukin-1 alpha (IL-1α), interleukin-1 beta (IL-1β), TNF-α, IL-6, and IL-8 all contribute to elevated body temperature (fever) and the production of acute-phase proteins.

Bear in mind that there are multiple innate sensors that function similarly to the PAMP-type molecules to allow for the recognition of pathogenic infection. For example, there are specific molecules to sense bacterial DNA, leading to the regulation of interferon production (STING; stimulator of interferon genes); other molecules have been identified that recognize RNA motifs. In many cases, these mediating sensors create an end outcome in which changes in the host cell surface molecule expression occurs. This serves as a direct link between initial innate immune responses and subsequent adaptive cell recognition of the invading pathogen.

MYCOBACTERIAL INFECTIONS

Mycobacterial infections such as tuberculosis and leprosy are extremely complex (Fig. 5.4). The mycobacteria have evolved to inhibit normal macrophage-killing mechanisms (e.g., phagosome-lysosome fusion) and

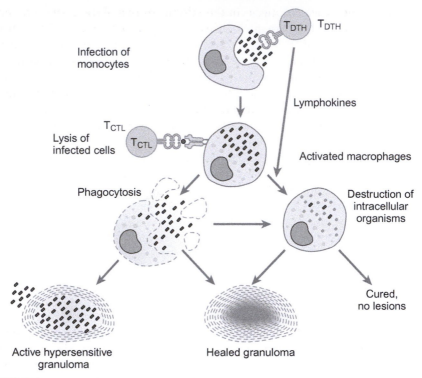

FIG. 5.4 Immune response to mycobacterial infections. Immunity against mycobacteria is initiated by phagocytic macrophages, the preferred host for the infectious agent. The overall outcome and associated pathologies depend upon the level of activation by cellular-delayed, type-hypersensitive (T_{DTH}) response.

survive within the "disarmed" professional phagocyte. T_{DTH}-initiated cellular responses are the main mechanisms involved in an active immune response, including granulomatous hypersensitivity, to wall off and contain organisms, but only after infection has become established. Both helper and cytotoxic T cells are responsible for controlling active infection, working through recognition of mycobacterial antigens and glycolipids released by infected cells. This leads to the release of cytokines and chemokines that recruit additional immune effectors.

A local environment is established to contain infection, resulting in granuloma pathology. Healing of the infected center may occur, with limited necrosis of the infected tissue. If the infection persists, an active caseous (cheesy-appearing) granuloma may form with a necrotic nidus comprised of infected and active macrophages. The granuloma is usually circumscribed by responding T cells, with host-mediated destructive response occurring inside the area of organism containment. The presence of giant cells (activated syncytial multinucleated epithelioid cells) is characteristic of a late-stage response.

At one time, it was thought that the tissue lesions of the disease tuberculosis required the effect of delayed hypersensitivity. The term "hypersensitivity" was coined because animals with cellular immune reactivity to tubercle bacilli developed greater tissue lesions after reinoculation of bacilli than did animals injected for the first time. The granulomatous lesions seen in tuberculosis depend upon primary innate functions, as well as acquired immune mechanisms; lesions are not the cause of disease but rather an unfortunate effect of protective mechanisms. This is readily apparent in secondary (or postprimary) disease, where a strong host immune component is required for the initiation of destructive pathology. In the lung, extensive damage with accompanying caseous granulomatous pathology can ultimately result in respiratory failure. The granulomatous immune response produces the lesion, but the mycobacterium causes the disease.

VIRAL INFECTIONS

Immune responses to viral agents depend upon the location of the virus within the host (Fig. 5.5). Antibodies play a critical role during the extracellular life cycle of the virus. Antibodies can bind to virus-forming complexes to inactivate virions and allow them to be cleared effectively by professional phagocytes. Humoral responses can prevent the entry of virus particles into cells by interfering with the ability of the virus to attach to a host cell, and secretory IgA can prevent the establishment of viral infections of mucous membranes. Once a viral infection is established within cells, it is no longer susceptible to the effects of antibodies. Upon entry to cells, immune resistance to viral infections is primarily T-cell mediated. To be effective in attacking intracellular organisms, an immune mechanism must have the capacity to react with cells in solid tissue. This is a property of cell-mediated reactions, especially cytotoxic T lymphocyte (CTL), but not of antibody-mediated reactions.

Most nucleated cells have an inherent, but limited, mechanism to downregulate viral replication through self-production of IFN-α and IFN-β. However, these nonspecific interferon responses are not sufficient to eliminate the virus. NK cells are an early component of the host response to viral infection. They nonspecifically recognize and kill virally infected targets, without use of an antigen receptor. In this way, the NK cell provides a link between innate and adaptive immunity as they produce multiple immunomodulatory cytokines [e.g., IFN-γ, TNF-α, transforming growth factor-β (TGF-β), and interleukin-10 (IL-10)]. In addition, NK cells release IFN-γ (which is physically different from the other interferons) and interleukin-12 (IL-12), molecules, which both activate macrophages and help to prime T cells for an effective antiviral T_H1 response.

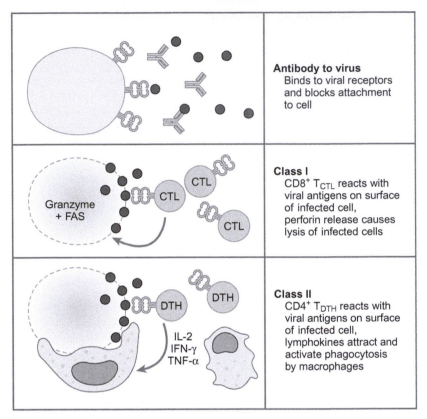

FIG. 5.5 Immune response to viral infections. Immunity against viral infections is three-fold, with contributions by antibodies, cytotoxic T cells, and T-helper cells.

At some stage of the infection process, viral-infected cells will express viral antigens on their cell surface in combination with class I molecules. Specific sensitized CD8+ CTL cells recognize presented viral antigens and destroy the virus-infected cells (and therefore limit viral replication) through the release of factors that include granzymes, perforins, and interferons. Lethal signals also may be delivered through Fas/Fas-ligand mediated mechanisms. Adverse effects occur if the cell expressing the viral antigens is important functionally, as is the case for certain viral infections of the central nervous system. If the virus-infected target is a macrophage, lymphocyte T-helper cells exhibiting delayed-type hypersensitivity (T_{DTH}) functions can activate the macrophages to kill their intracellular viruses; lymphokine-activated macrophages produce a variety of enzymes and cytokines that can inactivate viruses. The critical nature of T cell–mediated responses to viral infections is evident in patients who have defective CMI.

Human Immunodeficiency Virus

The human immunodeficiency virus-1 (HIV-1) is a member of the Lentivirus family and is the retrovirus most commonly associated with HIV infection in the United States and Europe. It has a high clinical importance and enormous social and economic impact throughout the world. As a retrovirus, it requires reverse transcription of its single-stranded RNA genome to a double-stranded DNA intermediate for integration into the host cell genome. The HIV virus infects CD4-positive cells, including T-helper lymphocytes, macrophages, and other cell types. The gp120 virus surface antigen binds with high affinity to the cell surface CD4, promoting fusion between virus and cell membranes. Coreceptor chemokine receptors CXCR4 and CCR5 also may play a role in internalization, as well as immunoglobulins that assist in antibody-dependent uptake. The natural history of HIV depicts the outcome of HIV infection, which leads to destruction of lymphocytes and development of acquired immune deficiency syndrome (AIDS) (Fig. 5.6).

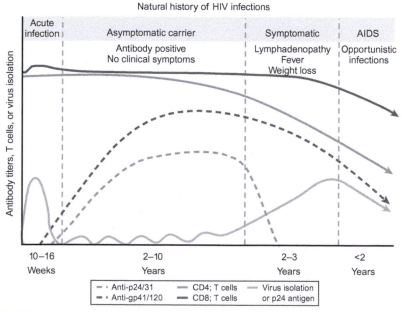

FIG. 5.6 The natural history of HIV infection. Infection of CD4+ lymphocytes, as well as other cell types, leads to virus production and cytolysis or long-term latent infection that progresses from primary infection through late symptomatic infection (AIDS). Accompanying this process are profound defects in T-helper and cytotoxic cell activity, with concomitant development of opportunistic infections.

PARASITIC INFECTIONS (HELMINTHS)

Host responses to parasitic worm infections are generally more complex because the pathogen is larger and cannot be engulfed by phagocytes (Fig. 5.7). The helminths typically undergo life-cycle changes as they adapt to life in the host. Worms are located in the intestinal tract, tissues, or both. Tapeworms, which exist only in the intestinal lumen, promote no protective immunologic response.

On the other hand, worms with larval forms that invade tissue typically stimulate an immune response. The tissue reaction to *Ascaris* and

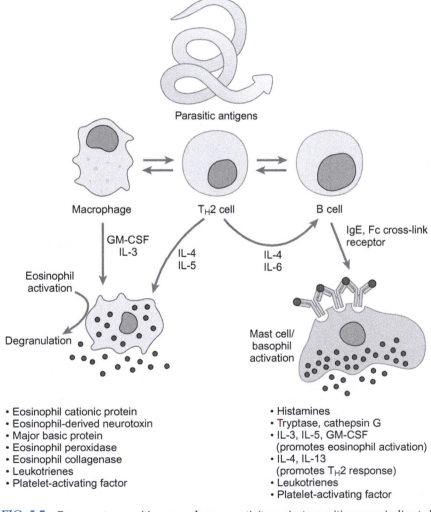

FIG. 5.7   Response to parasitic worms. Immune activity against parasitic worms is directed by T-helper 2 cells, driving the activation of eosinophils, basophils, and mast cells to release inflammatory mediators to limit parasitic activity and kill invading organisms.

Trichinella consists of an intense infiltration of polymorphonuclear leuko-cytes, with a predominance of eosinophils. Therefore, a variety of antigens that depend on the life-cycle stage are displayed in changing tissue environments. Numerous cells play a role, depending on the location of the organism. Antigens on the surface of organisms, or released into the local environment, may stimulate T cells and macrophages to interact with B cells to secrete specific antibodies. Interleukin-5 (IL-5), a T cell-derived factor, is instrumental in the stimulation of eosinophils, which act by associating with specific antibodies to kill worms by antibody-dependent cell-cytotoxic (ADCC) mechanisms or by releasing enzymes from granules to exert controlling effects on mast cells.

Antigen reacting with IgE antibodies bound to intestinal mast cells stimulates the release of inflammatory mediators, such as histamine, proteases, leukotrienes, prostaglandins, and serotonin. These agents cause an increase in the vascular permeability of the mucosa, exposing worms to serum immune components, and stimulate increased mucus production and increased peristalsis. These activities are associated with the expulsion of parasitic worms from the gastrointestinal (GI) tract through the formation of a physical barrier to limit adherence and interactions with the mucosal surface.

Eosinophil granules contain basic proteins that are toxic to worms. Eosinophils may be directed to attack helminths by cytophilic antibodies that bridge the eosinophil through the Fc region and the helminth by specific Fab-binding ADCC. Anaphylactic antibodies (IgE) are frequently associated with helminth infections, and intradermal injection of worm extracts elicits wheal-and-flare reactions. Children infested with *Ascaris lumbricoides* have attacks of urticaria, asthma, and other anaphylactic or atopic reactions presumably associated with the dissemination of *Ascaris* antigens.

FUNGAL INFECTIONS

Cellular immunity appears to be the most important immunologic factor in resistance to fungal infections, although humoral antibody certainly also plays a role in protection. T_H1 responses are protective via the release of IFN-γ. By contrast, T_H2 responses (IL-4 and IL-10) typically correlate with disease exacerbation and pathology. The importance of cellular reactions is indicated by the intense mononuclear infiltrate and granulomatous reactions that occur in tissues infected with fungi, and by the fact that fungal infections are frequently associated with depressed immune reactivity of the delayed hypersensitivity type (opportunistic infections). For example, the condition of chronic mucocutaneous candidiasis caused by persistent or recurrent infection by *Candida albicans* usually manifests only in patients with a general depression of cellular immune reactions. As a general rule, fungi appear to be resistant to the effects of antibodies, and CMI is needed for effective resistance.

EVASION OF IMMUNE RESPONSE

The ultimate evolution of the host-parasite relationship is not for a cure of an infection and complete elimination of the parasite; rather, it is a mutual coexistence without deleterious effects imparted to the host. In many human infections, the infectious agent is never fully destroyed and the disease enters a latent state. Infectious organisms have developed "ingenious" ways to avoid immune defense mechanisms. Organisms may locate in niches (privileged sites) not accessible to immune effector mechanisms (protective niches) or hide themselves by acquiring host molecules (masking). They may change surface antigens (antigenic modulation), hide within cells, and produce factors that inhibit the immune response (immunosuppression) or fool the immune system into responding with an ineffective effector mechanism (immune deviation).

Bacteria have evolved to evade different aspects of the phagocyte-mediated killing, as outlined in Fig. 5.8. Viral entities also subvert immune responses, usually through the presence of virally encoded proteins. Some of these proteins block effector functions of antibody binding, block

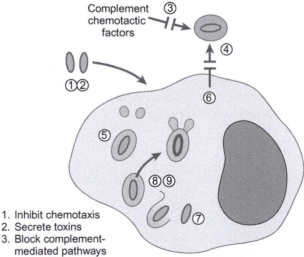

1. Inhibit chemotaxis
2. Secrete toxins
3. Block complement-mediated pathways
4. Have outer capsules to block attachment, phagocytosis
5. Inhibit lysosomal fusion
6. Have outer coat resistant to degrative enzymes
7. Escape from phagosome
8. Turn off cytokine activation
9. Activate cytokines inappropriately

FIG. 5.8 Mechanisms of infectious organisms to avoid immune defenses. Organisms evade effective immune responses through various mechanisms, including location in protective niches, acquisition of host molecules, alteration of surface antigens, and production of factors to inhibit or redirect the responses.

complement mediated pathways, inhibit the activation of infected cells, and can downregulate MHC class I antigens to escape CTL killing. The Herpes virus HSV-1 produces a factor that inhibits inflammatory responses by blocking the effects of cytokines through receptor mimicking. It also produces another protein capable of blocking that blocks antigen presentation and processing. Another example is the Epstein-Barr virus, which encodes a cytokine homolog of IL-10 to immunosuppress through the activation of T_H2 rather than T_H1 responses.

SUMMARY

- The immune response to initial infection is divided into phases. The first is an early innate and nonspecific response, where preformed effector cells and molecules recognize microorganisms. The next phase is again primarily a nonspecific encounter with the organism, characterized by recruitment of professional phagocytes and NK cells to the site of infection. The final phase involves antigen-specific effector cells (B and T lymphocytes) that undergo clonal expansion; these cells provide memory responses in case of reinfection.
- The host defense is based upon the availability of resources to combat a localized pathogen, contingent upon the life cycle of the pathogen and the extent of exposure to soluble factors or cellular processes.
- The immune mechanisms against classes of pathogens depend upon properties of the infectious agent or agents. The immune system has evolved to induce multiple arms of response to combat the broad array of organisms seeking to colonize the body.
- Virtually all classes of infectious agents have devised ways to avoid host defenses. These mechanisms include nonaccessibility in protective niches, antigenic modulation of surface molecules, and release of factors to either suppress the immune response or cause immune deviation and ineffective response to the pathogen.

Basic Disorders of Immune Function

CHAPTER FOCUS

To detail concepts associated with mechanisms underlying immunodeficiency, focusing on causes related to primary (genetic) components for clinical manifestation. Deficiencies in cell phenotypes [lymphocytes, natural killer (NK), and phagocytic cells], as well as innate components, will be discussed, along with immune-based treatment options for patients with congenital immunodeficiency. Finally, information will be presented regarding relative immunodeficiency as a predisposition for infection.

IMMUNODEFICIENCY DISORDERS

Immunodeficiency disorders are a diverse group of illnesses that result from one or more abnormalities of the immune system. The abnormalities can involve the absence or malfunction of blood cells (lymphocytes, granulocytes, monocytes) or soluble molecules (antibodies, complement components) that result from an inherited genetic trait (primary) or from an unrelated illness or treatment (secondary). The principal manifestation of immunodeficiency is an increased susceptibility to pathogens, as documented by increased frequency or severity of infection, prolonged duration of infection with development of an unexpected complication or unusual manifestation, or infection with organisms of low pathogenicity.

GENETIC BASIS FOR PRIMARY IMMUNODEFICIENCY

The altered genetic component gives rise to deficiencies in proteins and cellular functions. These are often defects in one particular component, leading to the disruption of a pathway that culminates in effective immune function. Fig. 6.1 depicts the general classes of immune deficiencies, subdividing defects in innate and adaptive mechanisms. The general mechanisms and associated clinical disorders are discussed in this chapter.

INNATE DEFICIENCIES

Deficiencies in innate components are at the heart of many immune disorders. Any functional defect in function of phagocytic cells, dendritic cells, neutrophils, or natural killer (NK) cells results in the inability to control invading pathogens rapidly. Likewise, defects in the proteins involved in early pathways of the complement cascade limit subsequent events that control postinfection responses.

Immunodeficiency		
B-cell deficiencies	**T-cell deficiencies**	
Recurrent bacterial infections	*Severe viral, fungal, and protozoal infections*	
Bruton's agammaglobulinemia—defect in B-cell development Common variable-hypogammaglobulinemia—defect in plasma cell differentiation Hyper-IgM syndrome—defect in class switching	**B- and T-cell deficiencies** Severe combined immunodeficiency	Bare lymphocyte syndrome—lack of class II MHC Omenn syndrome—defect in TCR gene rearrangement DiGeorge syndrome—thymic aplasia
Phagocytic cell deficiencies	**Complement deficiencies**	
Recurrent bacterial infections	*Recurrent bacterial infections Defects in immunocomplex clearance*	
Chronic granulomatous disease—lack of respiratory burst Leukocyte adhesion deficiency—lack of PMN extravasation into tissue Chediak-Higashi syndrome—defect in neutrophil microtubule function and related phagosome/lysosome fusion	C1, C2, or C4 deficiency—defects in clearing immunocomplexes C3 or C5 deficiency—block in alternative and classical pathways C6, C7, C8, or C9—defect in MAC assembly and function	

FIG. 6.1 Primary immunodeficiencies. Manifestation of immunodeficiency depends on the etiology of response. B-cell deficiency is marked by recurrent infections with encapsulated bacteria. T-cell deficiency manifests as recurrent viral, fungal, or protozoal infection. Phagocytic deficiency, with an associated inability to engulf and destroy pathogens, usually appears with recurrent bacterial infections. Complement disorders demonstrate defects in activation patterns of the classical, alternative, and/or lectin-binding pathways and related host defense mechanisms.

Chronic Granulomatous Disease

Chronic granulomatous disease (CGD) represents a defect in phago-cytic cells normally associated with engulfment and subsequent respiratory oxidative activity, superoxide production, and hydrogen peroxide creation that is essential for killing phagocytosed bacteria. Leukocytes from an individual with CGD demonstrate defects due to genetic abnormalities that limit intracellular enzymatic function and effectiveness of the phagolysosome. In many cases, there is no dysfunction in chemotactic response, and it is typical to observe high leukocyte counts in people with this disorder. However, bactericidal functions are defective, and increased frequency of cellular response is ineffective. Of note, very effective laboratory tests are available to diagnose CGD, including the nitroblue tetrazolium dye test, the dihydrorhodamine test, and the chemiluminescence assay, all of which measure levels of hydrogen peroxide and subsequent superoxide production.

Leukocyte Adhesion Deficiency

Leukocyte adhesion deficiency (LADs) is an autosomal recessive disorder characterized by recurrent bacterial and fungal infections. In these patients, leukocyte adhesion-dependent functions are impaired or absent. The molecular basis of the defect is absent or deficient expression of the glycoprotein β2 integrins CD11 or CD18, which participate in leukocyte adhesion to cells and parenchyma. Another variant of the LAD syndrome occurs when carbohydrate ligands on neutrophils are defective for binding to cytokine-activated endothelium, which inhibits neutrophil movement into tissue. Of interest, these individuals usually display impaired wound healing as well.

Chediak-Higashi Syndrome

Chediak-Higashi syndrome reflects a defect in neutrophilic microtubule function. Essentially, this dysfunction presents through an inability of the lysosome and phagosome to fuse after ingestion of microorganisms. Lysosomal trafficking is impaired, resulting in high incidence of recurrent pyogenic infections.

COMPLEMENT DISORDERS

The complement system comprises distinct serum proteins that effect multiple biological functions. In addition to direct bactericidal function, the proteins and their enzymatically produced products regulate cellular chemotaxis, phagocytic functions through opsonization, vascular modulation, and direct activation of multiple cell phenotypes. Genetic deficiencies or mutations of any complement cascade zymogen lead to impaired host defense.

Complement factor C3 deficiency is a rare disorder exemplifying the serious nature of complement dysfunction. C3 is the pivotal complement component for classical and alternative pathway activity. Without C3, the anaphylatoxins C3a and C5a are not released and phagocytic cells are not chemotactically driven to the site of infection. The result is that bacteria are not effectively opsonized, and polymorphonuclear leukocytes and monocytes are not driven to increase phagocytosis. General dysfunction of the C5–C9 membrane attack complex is impaired. Individuals are especially susceptible to pyogenic infections, as well as to meningococcal and gonococcal infections. **Hereditary angioedema** reflects another specific disorder of the complement cascade, related to pathways that control C1 turnover. In this case, continued production of C1 generates excess vasoactive mediators C3a and C5a, causing capillary permeability and edematous activity.

Innate Pattern-Recognition Receptor (PRR) Disorders

Toll-like receptors (TLRs) represent innate receptors that regulate responses to motifs found on pathogenic invaders. As such, TLRs initiate the first line of defense to trigger cytokine responses that dictate effective lymphocytic involvement. Although much is still to be understood regarding this subclass of pattern-recognition receptors (PRRs), it is clear that absence or defects in recognition of pathogen-associated molecular patterns (PAMPs) impair the recruitment of leukocytes to sites of infection and reduce the natural augmentation of antimicrobial or antiviral activity. Furthermore, lack of engagement of the PRRs renders dendritic cells defective in their ability to migrate to draining lymph nodes, where they effectively present antigen to T cells.

ADAPTIVE IMMUNE DISORDERS

Genetic abnormalities are also at the base of the lymphocyte immunodeficiencies. Both B and T lymphocytes begin their journey in the bone marrow; any underlying genetic abnormality caused by absent or altered enzymes can arrest maturation at defined stages of hematopoietic development. Improper genetic control for rearrangements of the antigen receptor genes will result in a lack of the mature cells required for effective adaptive immune function.

X-Linked Agammaglobulinemia (X-LA)

By virtue of their lack of mature B cells, patients with X-linked agam-maglobulinemia (X-LA), also known as Bruton agammaglobulinemia, have no circulating plasma cells to secrete immunoglobulins. As such, these individuals are extremely susceptible to infection with multiple classes of pathogenic organisms. The molecular basis has been linked directly to a defective cytoplasmic tyrosine kinase, which prevents B-lymphocyte maturation and the eventual production of immunoglobulins. Recurrent bacterial and pyogenic infections begin early in life (between 6 and 9 months of age) when maternal antibodies decrease due to natural decay. Individuals diagnosed with X-LA are susceptible to infections including agents that cause pneumonia, sinusitis, otitis, and meningitis. Sepsis may also result from infection with capsular-coated organisms, such as *Pneumococci* and *Streptococci*, which require immunoglobulin G (IgG) isotype antibodies for opsonization and subsequent phagocytosis for targeted destruction.

Selective (IgA) Deficiency

Selective immunoglobulin A (IgA) deficiency is the most common form of immunodeficiency. As the name indicates, individuals with this deficiency lack IgA, which is a critical isotype to protect against infections of mucous membranes lining the mouth, airways, and digestive tract. Gastrointestinal (GI) infection and malabsorption issues are typically prominent, which is expected because IgA coats mucosal surfaces. The clinical definition includes undetectable serum IgA in the presence of normal serum levels of IgG and immunoglobulin M (IgM). The mechanisms for the failure of IgA to be secreted are in genetic lesions; in many cases, the alpha-heavy chain gene remains intact, and yet a problem remains with both IgA production and secretion.

Common Variable Immunodeficiency

Patients with common variable immunodeficiency possess malfunctioned B cells that demonstrate a developmental block in plasma cell differentiation. This results in an essential failure to secrete immunoglobulins. They also lack effective T-helper-cell function. The group is typically diagnosed between the ages of 15 and 35 years, when they begin to show signs of recurrent bacterial infections. Concurrently, there is a decreased immunoglobulin load and associated impairment in humoral responses. Cellular immunity is usually normal. The defects have been linked to proteins involved in the activation and regulation of isotype switching.

Clinical presentation includes persistent lung infections, as well as the presence of intestinal pathogens that are cleared normally by healthy individuals. Upon immunization, only low levels of IgM isotype antibodies are produced, with no IgG produced even after multiple immunizations.

Hyper IgM (HIM) Disorders

Hyper IgM (HIM) is an immunodeficiency in which high levels of circulating IgM are present at the expense of other antibody isotypes. In males, the abnormal gene in the X-linked type (HIM-1) is due to abnormal production of a required ligand (CD40L) present on T cells that modulates interaction between the cognate CD40 ligand on B cells. This interaction is required for cellular activation, as well as for proper isotype-switching events. It should be noted that females may also exhibit a form of HIM termed HIM-2, which correlates with defective activation-induced deaminase or uracil DNA glycosylase.

DiGeorge Syndrome

DiGeorge syndrome represents a severe T-cell deficiency; it is characterized clinically by an absent thymus, as well as by associated developmental abnormalities in the newborn. As expected with defective T cells, there is an increased susceptibility to opportunistic infections, most notably fungal infections in the very young. Infants are typically seen with normal numbers of B lymphocytes and the presence of serum immunoglobulins. However, these infants fail to mount effective antibody responses due to the lack of T-lymphocyte helper-cell activity.

Wiskott-Aldrich Syndrome

Wiskott-Aldrich syndrome (WAS) is a rare X-linked disorder where mutations in the key regulator of actin polymerization (WAS protein) have been identified. Mutations give rise to deficiencies in signaling, cell locomotion, and immune synapse formation. The immune deficiency leads to decreased antibody production and the inability of T cells to become polarized, thus allowing diagnostic placement as a combined immunodeficiency.

Severe Combined Immunodeficiency (SCID)

There are many types of combined lymphocytic disorders in which deficiency in both B- and T-cell populations result in susceptibility to recurrent life-threatening infections. The classically defined form of severe combined immunodeficiency (SCID) is X-linked, but autosomal recessive forms also exist. Phenotypic analysis reveals an absence of lymphocytes that bear mature cell surface molecules, functional antigen receptors, or both. Often, lymphocytes also exhibit an inability to respond to mitogenic stimulation. Frequently, there are small numbers of B cells and few serum immunoglobulins. The molecular defect in X-linked SCID is a mutation in the gene that codes for the gamma chain of the interleukin-2 (IL-2) receptor; this chain shares functions with multiple cytokine receptors, thus impairing T-cell maturation and proliferation and basically eliminating T-cell functionality. The SCID defect occurs because the defective interleukin-2 gamma (IL-2γ) chain renders dysfunctional not only the IL-2 receptor, but also interleukin-4 (IL-4), interleukin-7 (IL-7), interleukin-15 (IL-15), and interleukin-21 (IL-21).

There are several other forms of SCID involving genetic lesions (all autosomally inherited) in multiple critical genes needed for adaptive immune function. For example, **Omenn syndrome** is an autosomal-recessive SCID in which the recombination-activating genes (*RAG1* and *RAG2*) involved in antigen receptor DNA rearrangement render that protein dysfunctional, adversely affecting the circulating levels and functionality of both B and T cells. Two other disorders that represent T-cell deficiencies should be included in this overall discussion. **Ataxia telangiectasia** is an autosomal-recessive disorder in which thymic dysfunction does not permit T-cell development. Low numbers of both CD4+ and CD8+ cells ensue. **Bare lymphocyte syndrome** has a mechanism directly affecting CD4+ cells; the lack of Class II histocompatability molecules prevents positive selection of the T-helper-cell phenotype during maturation in the thymus.

TREATMENT OF IMMUNODEFICIENCY DISEASES

The effective treatment of immunodeficiency is in large part dependent upon identification of the underlying disorder. A major challenge presented by primary immunodeficiency is to translate genetic and molecular discoveries into new therapies for patients. In general, all disorders require general supportive care, with constant guard against infectious assault. In many cases, a bone marrow transplant is an effective means of replacing hematopoietic stem cells, although complications may arise due to transplantation-related issues. In essence, a bone marrow transplant entails a procedure in which alloreactive hematopoetic stem cells are given to recipients so they can adopt a new set of lymphoid and myeloid progenitors. Significant advances have been made to increase the success of stem cell transplantation, targeting cell preparation and manipulation, selection of donor cell populations, and the use of chemotherapeutic agents. Another type of treatment relies on cytokine therapy, which can be particularly effective in selected defects where specific lack of a component is identified. The same holds true for complement defects; if the missing factor is known, it can be therapeutically administered to recover function.

Clinical success has been seen with many other palliative treatments. B-lymphocyte disorders are primarily treated with **intravenous immunoglobulin (IVIG)**. This mechanism of passive antibody transfer restores regular immune function, although continued administration is required every 3–4 weeks due to the half-life of the IgG isotype given. IVIG is also effective to treat disorders of other cell phenotypes that ultimately affect antibody production. Regarding phagocytic disorders, supplementation with cytokines such as interferon-gamma may be highly effective, especially in the case of chronic granulomatous disorders. Additional immunotherapies using biologics, such as monoclonal antibodies, that may be beneficial for individuals with immune disorders will be discussed in Chapter 11.

Genomic screening of individuals with immune disorders has greatly advanced our ability to pinpoint genetic deficiencies underlying many immune system dysfunctions. This field is a fast-growing one, with major recent advancements made in defining the nucleotide sequences responsible for clinical phenotypes. Techniques such as next-generation sequencing allow exact DNA identification of causative nuclear regions, replacing earlier (and powerful) gene-mapping and candidate-gene analyses. This generates excitement that mutations can be screened and therapies put in place to target altered gene products so as to recover immune function. Indeed, there may be future opportunities, perhaps using the **CRISPR-cas9** technology, a way to introduce targeted changes in the genome, to create corrective therapies.

While physicians, not to mention the population at large, grapple with the ethics underlying new technologies for diagnosis and repair, there remains a common thread of hope that physicians will adopt a universal neonatal screening process to at least identify patients with primary immunodeficiency. This would allow early diagnosis and effective treatment prior to presentation with life-threatening infections or the establishment of cancer due to a lack of natural protective immune-surveillance.

IMMUNODEFICIENCY AS A PREDISPOSITION TO DISEASE

Immunodeficiency is closely associated with expected changes due to lack of the functional cells required for the common surveillance functions of the host. As such, there is a high incidence of both cancer (viral-induced) and autoimmune dysregulation over time, especially in individuals who show defects in lymphocytic populations. However, the principal manifestation of immunodeficiency is an increased susceptibility to infection, as documented by increased frequency or severity of infection, prolonged duration of infection with development of an unexpected complication or unusual manifestation, or infection with organisms of low pathogenicity.

Events of stress affect immune function. Certainly, acute stress elicits physiological responses that prime or enhance immunity in preparation for injury or an infectious threat. However, chronic stress and distress (on the order of weeks, months, or years) result in immune dysregulation and direct immunosuppression. In chronic stress, it has been shown that the physiological response persists long after cessation of the stressor (changes in diet, sleep, exercise, environment, or neuroendocrine reactivity). Shifts in T-helper-cell function occur in situations of prolonged stress, predominantly controlled by neuroendocrine factors regulated by the hypothalamus-pituitary-adrenal axis. Thus, we become more susceptible to infectious agents. Psychological and emotional stressors, in addition to factors regulating physiological parameters, are included in this regulation.

Human Immunodeficiency Virus (HIV) and Acquired Immunodeficiency Syndrome (AIDS)

The interaction between the **human immunodeficiency virus (HIV)** and the immune system is complex and represents a secondary immunodeficiency related to the viral destruction of lymphocytes. The manifestation of the infection results in **acquired immunodeficiency syndrome (AIDS)**, which is directly related to decreased CD4+ T-cell numbers and function (as reviewed in Chapter 5). In turn, this increases susceptibility to infection by intracellular pathogens, viruses, and fungi. It also causes a decrease in surveillance activity against certain tumors. Because CD4+ T cells play a central role as helper cells and mediators of delayed-type hypersensitivity, subsequent loss of this population will leave the individual exposed to attack by many opportunistic agents.

SUMMARY

- Inherited gene defects are primary causes of immunodeficiency. Defective function may occur in both innate and adaptive immune-system cells.
- The lack of immune effector functions generally results in increased incidence of infection. However, lack of regular immune surveillance also can lead to increases in cancer occurrence, as well as autoimmune dysfunction.
- Diagnostic tests augment the medical history and physical exam. A process to screen newborns for primary immunodeficiencies would allow early diagnosis and effective treatments prior to presentation with life-threatening infection.
- Therapeutic intervention depends on the nature of the immune deficiency, with the most successful treatments relying on corrective replacement of defective or absent immune functions.

Autoimmunity: Regulation of Response to Self

CHAPTER FOCUS

To discuss concepts associated with mechanisms underlying the development of autoimmunity. Elements underlying autoimmune dysfunction will be examined, moving from basic concepts of tolerance to specific mechanisms involved in major clinical disorders when tolerance to self-antigens is lost. The goals are to present the development of autoimmune reactivity so that clinical disease and outward symptoms are understood as being related to underlying immune mechanisms, and to identify molecular targets involved in the host's self-recognition response. The discussion also will include therapeutics that target immune parameters, as well as laboratory tests for specific autoimmune disorders.

HOMEOSTASIS, IMMUNE REGULATION, AND AUTOIMMUNITY

The regulation of immune function and overall immune homeostasis is under the control of multiple factors that include genetic and environmental components. Human leukotype antigen (HLA) allotypes, antigen dose, and existing cytokine milieu can all influence responses to commonly encountered antigens. **Autoimmunity** represents the manifestation of a specific adaptive response to self-antigens. Paul Ehrlich (1854–1915) realized that reactivity toward one's own self-tissue was a pathologically destructive process, which he referred to as "horror autotoxicus."[1]

[1]Ehrlich, P. (1957). Die Schutzstoffe des Blutes. In: Himmelweit, F. (Ed.),The Collected Papers of Paul Ehrlich, Vol.2. London: Pergamon Press, pp. 298–315.

We now realize that the autoimmune process is a complex interaction where specific adaptive responses are mounted against self-antigens due to the loss or circumvention of tolerance-related mechanisms. Auto-immune diseases, therefore, result from the dysregulation of immune pro-cesses and pathways that are involved in normal immune function, with resulting pathological damage to self-tissue.

TOLERANCE TO SELF

At the basis of autoimmunity is the loss of control of self-reactive T lymphocytes (Fig. 7.1). Induction of **tolerance** in immature lymphocytes is critical for the elimination of self-reactive cells. Tolerance begins during development. **Central tolerance** is provided during lymphocytic matura-tion and involves the physical destruction of self-reactive cells. For T lymphocytes, negative selection via apoptosis in the thymus eliminates the majority of cells that bear high reactivity to self-peptides that can be accommodated into the presenting grooves of the HLA molecules. The elimination of self-reactive T cells involves a series of interactive

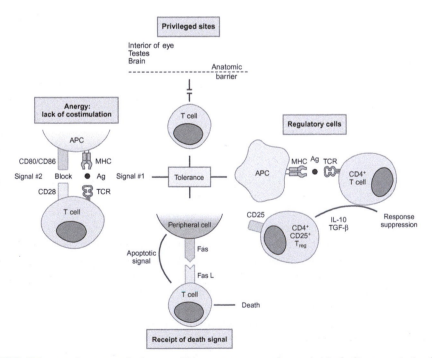

FIG. 7.1 Mechanisms of tolerance. Tolerance can occur by means including apoptosis of reactive cells, development of anergic response to antigen through loss of secondary signals, regulation of response as a result of antigen excess, and active suppression by regulatory T cells.

events with various cell phenotypes throughout compartments in the thymus. Basically, there is a balance in the relative strength of interactions with HLA molecules, which naturally express self-peptides (in the absence of infection). B lymphocytes also undergo central tolerance in a more complicated scenario, which takes place during developmental stages in the bone marrow.

However, it is now well recognized that it is nearly impossible to eliminate all reactive lymphocytes in the body completely. **Peripheral tolerance** in healthy individuals is a way to ensure that cells do not react to self-antigens in the secondary lymphoid tissues. Mature cells leaving the thymus may be tolerized, subject to antigenic factors that affect immunogenicity. Factors that induce tolerance under normal conditions include extremely high- or low-dose antigens, weak binding to histocompatability molecules, and route of administration (e.g., oral antigens are well tolerated). Peripheral tolerance occurs primarily via **anergic** reactions. Induction of "normal" T-cell reactivity requires multiple signals, including antigen recognition by the antigen receptor and secondary signals from adhesion and costimulatory molecules on the cell surface. If specific costimulatory signals (CD80 or CD86) on the surface of the presenting cell are absent, then cognate molecules (CD28) on the T cell are not engaged, which leads to a nonreactive, anergic state. Another mechanism, referred to as an **immune checkpoint**, utilizes surface molecules on T cells (CTLA-4, LAG-3, PD-1) designed to regulate immune activation; these molecules play key roles in preventing reactivity to self. In addition, a population of CD4+ T cells, called **T-regulatory cells**, or **Tregs**, can actively anergize lymphocytes in the periphery that escape primary selection. This is done through the focused release of cytokines that downregulate activities. Realize that the regulation of self-reactive T-cell activity is under constant control, and there is oversight of an exquisite balance of cytokine signals, costimulatory surface molecules, and environmental cues. Of further interest, subpopulations of CD8+ Treg cells have been identified recently; the potential for CD8+ cells to regulate tolerance is an area of active investigation.

Tolerance may also occur in immunologically privileged sites, such as the eye, brain, and testes, usually due to local secretion of immunosuppressive factors (e.g., TGF-β). It is of particular interest that if hidden ("cryptic") antigens are exposed during trauma or tissue damage, they now can potentially serve to initiate responses to overcome toleragenic states. The developing fetus in the uterus is a unique example of a privileged site where the recognition of foreign antigens by maternally reactive lymphocytes is suppressed.

Immature B cells in the bone marrow undergo apoptosis upon the binding of self-antigens by activation-induced cell death mechanisms. Alternatively, the B cell may undergo receptor editing to change the binding specificity of the surface immunoglobulin, thus rendering the cell no longer self-reactive. Once in the peripheral tissue, B cells may undergo transition to a nonresponsive anergic state, a process which is dependent upon the level of specific antigens. Low-dose, soluble monomeric antigens may also induce tolerance, as they do not permit receptor cross-linking on the surface of the cell; this process results in clonal inactivation of the B lymphocyte. Excessively high antigen dosage can also result in anergic response due to overwhelming recognition in the absence of sufficient T-cell costimulation.

Finally, self-reactive B cells that escape the elimination or induction of anergy may be incapable of activation due to the lack of available T cells to help initiate the development of autoimmune responses. Although at times, T cells can later become activated to bacterial antigens with cross-reactivity to self-molecules (called **molecular mimicry**), when this occurs, naturally existing tolerization states may be negated.

ETIOLOGY OF AUTOIMMUNE DISEASE

It is clear that a combination of environmental and genetic components presents risk factors for autoimmune disease. Genetic factors have been identified that implicate polymorphisms in cytokine genes or their receptors, defective apoptosis genes, and complement component deficiencies. Infections or exogenous agents that cause physical damage are also likely to play important roles. The mechanisms underlying all autoimmune diseases are not fully elucidated; however, polymorphisms of MHC class II genes (alleles of HLA-DR, HLA-DQ, or both) are associated with increased susceptibility to autoimmune diseases. Indeed, clinical assessment has shown a high incidence in autoimmunity in individuals possessing the HLA-DR4, HLA-DR8, and DQ2 alleles. Possible mechanisms for a loss of tolerance include (1) a lack of Fas-Fas ligand-mediated deletion of autoreactive T cells in the thymus during development, (2) loss of T-regulatory or T-cell cytokine-mediated suppressor function, (3) cross-reactivity between exogenous and self-antigens (molecular mimicry), (4) excessive B-cell function due to polyclonal activation by exogenous factors (viral or bacterial origin), (5) abnormal expression of MHC class II molecules by cells that normally do not express these surface molecules, and (6) release of sequestered self-antigens from privileged sites, thus priming for responses to antigens not previously seen by the immune system.

As a rule, autoimmune disease symptoms vary greatly among individuals. Periods of extreme reactivity are often interspersed with asymptomatic periods of remission. Initial autoimmunity symptoms often include fatigue, rash, general or localized pain, and low-grade fever. A classic indicator of autoimmunity is inflammation, represented as immune reactivity present at both the local and systemic level.

Autoimmune diseases can be classified as organ specific or systemic in nature (Table 7.1). Three major types of mechanisms are recognized as causing different autoimmune disorders (Fig. 7.2). Two of these mechanisms involve autoantibodies directed against self-antigens; for both, classical complement pathway activation exacerbates local damage and inflammatory responses. In the first case, autoantibodies may be directed against a specific self-component, such as a surface molecule. Examples include antibodies against the acetylcholine receptor producing myasthenia gravis or antithyroid-stimulating hormone receptor antibodies that cause Graves' disease. The second mechanism involves autoantibodies which bind with antigens present in the blood, forming antigen-antibody (immune) complexes that deposit in organs or vascular beds, thus inciting an inflammatory response. This occurs in lupus glomerulonephritis, where complexes of anti-DNA antibodies and bound DNA accumulate in the kidney.

TABLE 7.1 Autoimmunity and Disease

Autoimmune disease	Mechanism	Pathology
Autoimmune hemolytic anemia	Autoantibodies to RBC antigens	Lysis of RBCs and anemia
Autoimmune thrombocytopenia purpura	Autoantibodies to platelet integrin	Bleeding, abnormal platelet function
Myasthenia gravis	Autoantibodies to acetylcholine receptor in neuromuscular junction	Blockage of neuromuscular junction transmission and muscle weakness
Graves' disease	Autoantibodies to receptor for thyroid-stimulating hormone	Stimulation of increased release of thyroid hormone (hyperthyroidism)
Hashimoto's thyroiditis	Autoantibodies and autoreactive T cells to thyroglobulin and thyroid microsomal antigens	Destruction of thyroid gland (hypothyroidism)
Type I diabetes (insulin-dependent diabetes mellitus)	Autoantibodies and autoreactive T cells to pancreatic islet cells	Destruction of islet cells and failure of insulin production
Goodpasture's syndrome	Autoantibodies to type IV collagen	Glomerulonephritis
Rheumatic fever	Autoantibodies to cardiac myosin (cross-reactive to streptococcal cell wall component)	Myocarditis

Continued

TABLE 7.1 Autoimmunity and Disease—cont'd

Autoimmune disease	Mechanism	Pathology
Pemphigus vulgaris	Autoantibodies to epidermal components (cadherin, desmoglein)	Acantholytic dermatosis, skin blistering
Multiple sclerosis	T-cell response against myelin basic protein	Demyelination, marked by patches of hardened tissue in the brain or the spinal cord; partial or complete paralysis and jerking muscle tremors
Systemic lupus erythematosus	Circulating immunocomplexes deposited in skin, kidneys, etc., formed by autoantibodies to nuclear antigens (antinuclear antibodies), including anti-DNA	Glomerulitis, arthritis, vasculitis, rash
Rheumatoid arthritis	Autoantibodies to IgG (rheumatoid factors); deposition of immunocomplexes in synovium of joints and elsewhere; infiltrating autoreactive T cells in synovium	Joint inflammation, destruction of cartilage and bone
Celiac disease	Antibodies made to gliadin (gluten), cross-reactive to tissue transglutaminase	Gluten-sensitive enteropathy, villous destruction, and gastrointestinal manifestations
Scleroderma (systemic sclerosis)	Antibodies to topoisomerases, polymerases, and fibrillarin	Skin-related fibrosis, damage to related arteries
Ankylosing spondylitis	CD4+ cells, possible activity to self-antigens (arthritogenic peptides, molecular mimicry, or aberrant forms of B27)	Rheumatic disease of joints and spine
Sjögren's syndrome	CD4+ cells, possible activity to self-antigens (M3 muscarinic acetylcholine receptor)	Lymphocytic-mediated destruction of lachrymal and salivary glands

The third mechanism involved in autoimmune activity is that of autoreactive T cells that recognize organ-specific self-antigens, leading to direct tissue damage. A specific subset of T helper cells, the T_H17 population, produces IL-17 within tissue microenvironments under chronic inflammatory conditions. IL-17 has been shown to play a critical role in pathogenesis of autoimmune inflammatory diseases, likely by

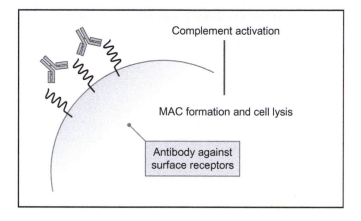

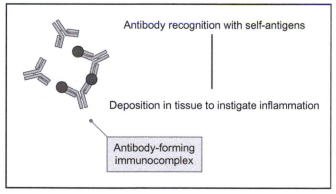

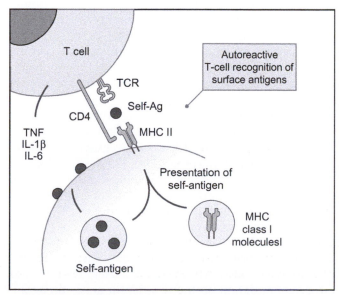

FIG. 7.2 Major mechanisms of autoimmunity and disease. *MAC*, membrane attack complex; *IL-1 β*, interleukin-1 beta.

dysregulating the delicate balance between self-reactive T cells and Tregs. The overabundant activity of T_H17 and the presence of IL-17 have been shown to coincide with diseases such as multiple sclerosis and rheumatoid arthritis, discussed next.

ROLE OF AUTOANTIBODIES AND SELF-REACTIVE T LYMPHOCYTES IN AUTOIMMUNE DISORDERS

The major autoimmune disorders and accompanying clinical presentations are listed in Table 7.1. What is critical to realize is that clinical presentation of autoimmune disorders is the culmination of multiple stages of misdirected or dysregulated immune function. Historically, autoimmune diseases were characterized solely by the presence of autoantibodies, or the presence of T cells reactive to identified antigens. However, researchers have come to appreciate that the architecture of the disorder is based on a combination of both innate and adaptive components. So while there may be a predominately dysfunctional arm of immune reactivity, it is more likely that multiple effector pathways and cell phenotypes are affected. Finally, it should be emphasized that in many cases, autoreactive T cells coexist with autoantibody responses, leading to exacerbation of disease and organ damage. The more common autoimmune disorders are discussed next.

Autoimmune hemolytic anemia is characterized by autoantibodies against antigens found on red blood cells (RBCs). Two types of antibodies exist. The "warm-reactive" autoantibodies are of the immunoglobulin G (IgG) isotype and react with the rhesus antigens (Rh antigens) on the cell surface at body temperature. This results in RBC opsonization and subsequent macrophage phagocytosis. There are also cold-reactive immunoglobulin M (IgM) antibodies that react with a different surface antigen (the I antigen), activating complement and mediating lytic events at lower temperatures.

Myasthenia gravis is a disorder where autoantibodies against the alpha chain of the nicotinic acetylcholine receptor in the neuromuscular junction act as an antagonist, resulting in dysregulation of muscle activity. This leads to clinical symptoms of muscle weakness, diplopia, dysarthria, and dysphagia. Of interest, because the isotype of the immunoglobulin is an IgG, it is possible for maternal antibodies to be transferred across the placenta to generate symptoms in the newborn.

Graves' disease patients demonstrate symptoms of hyperthyroidism, which include exacerbated weight loss, increased metabolism, palpitations, and fatigue. Classic physical symptoms include ophthalmopathy and changes in eye socket orbit. In this disorder, autoantibodies against the thyrotropin stimulating hormone receptor act as an agonist, leading to thyroid overactivation. As discussed with myasthenia gravis, maternal antibodies can be transmitted to the fetus, resulting in transient neonatal hyperthyroidism.

Systemic lupus erythematosus is another disorder that is primarily autoantibody in origin. This disease is characterized by systemic autoimmunity and multiorgan involvement. In this case, there is high production of autoantibodies against nuclear components. Reactivity during tissue damage leads to the creation of circulating immune complexes that deposit in tissues (skin, joints, kidneys) and heighten the development of hypersensitive pathology, leading to excess inflammation and chronic tissue damage.

Hashimoto's thyroiditis is a mixed disorder that is immune-characterized by the presence of both autoantibodies and autoreactive T cells to thyroglobulin and thyroid microsomal antigens. Autodestruction of the thyroid gland leads to hypothyroidism and associated symptoms (fatigue, goiter, weight gain). T-helper type 1 cells are involved in disease manifestation.

Celiac disease is defined by classical mucosal change of the small intestine, triggered by gluten in the diet. The patient exhibits a failure to thrive, with constant diarrhea, weight loss, and deficiencies due to inadequate nutrient uptake. Antibody reacts to gliadin, a protein found in wheat. However, the pathology is due to cross-reactive destructive response to self-transglutaminase and endomysium proteins.

Rheumatoid arthritis is a mixed-phenotypic autoimmune disorder. It was historically identified by the presence of **rheumatoid factor**, which is a group of antibodies directed against the constant portion of the IgG isotype molecule. More recently, it has been shown that antibodies are also made to citrullinated peptides, but it is as yet unknown if this is diagnostic or mechanistic. T lymphocytes are involved in clinical manifestation of symptoms and are readily identifiable as infiltrates in synovial spaces. Overall, the immune complex formation and associated T-cell activation lead to activation of innate components. Subsequent synovial inflammation ensues, culminating in the destruction of cartilage and development of bone erosion. Similar mechanisms of autoreactivity are also known to occur in related ankylosing spondyloarthritis.

Multiple sclerosis is an immune-mediated disease in genetically susceptible individuals where demyelination and axonal injury and destruction lead to slower nerve conduction. This leads to neurological dysfunction. The heart of the dysregulation is based in a T lymphocyte-mediated response where demyelination is accompanied by inflammation. Lesions throughout the brain and spinal cord continually appear and heal over time. The possible mechanisms associated with axonal loss during disease states are associated with the presence of activated CD4+ T cells targeted to myelin proteins. It is thought that they migrate to the central nervous system and assist in the activation of macrophages and B cells. This culminates in the secretion of proinflammatory cytokines, as well as antibodies, that continue to exacerbate the degenerative process.

Type I diabetes mellitus is characterized in a manner different from the disorders listed previously. In this case, autoreactive CD8+ cytotoxic T lymphocytes are found to be reactive to pancreatic islet cells. Subsequent targeted islet cell destruction leads to a failure of insulin production. Of note, although autoantibodies to insulin and islet cell antigens also may be present during disease states, they are probably diagnostic rather than causative of the disorder.

Sjögren's syndrome is a chronic disorder in which leukocytes destroy the exocrine glands, specifically the salivary and lacrimal glands that produce saliva and tears. Although the exact antigenic targets affected remain unknown, the high prevalence of disorder in specific HLA populations indicates that the pathology is due to a mixed response, whereby T lymphocytes provide assistance in the induction of self-reactive antibodies. The pathology may be secondary to an underlying connective tissue disease.

LABORATORY TESTS FOR AUTOIMMUNITY

The diagnostic criteria for autoimmunity is based on the presentation of symptoms, usually with patient presentation of at least one clinical impairment (or a series) that can be identified along a graded scale. Many times, the evaluation of disorder is accompanied by the presence of a positive laboratory test, with the persistence of immune markers present over multiple time periods (Table 7.2). The majority of diagnostic tests rely on the presence of autoantibodies. As such, many tests overlap with common autoreactive antigens, thus limiting their use as a sole diagnostic identifier for any particular disorder. For example, autoantibodies against nuclear components are found in multiple diseases. As such, they may be useful in identification of disorder, but only when combined with clinical criteria, including the presence of serum proteins containing acute phase reactants (C-reactive proteins, or CRPs) and proinflammatory cytokines [interleukin-1 (IL-1), interleukin-6 (IL-6), and tumor necrosis factor alpha (TNF-α)]. Serum levels of complement components also can serve as markers of disease activity, with a specific decrease in C3 levels consistent with complement consumption due to autoantibody activity. Regarding specific T cell responses, a challenge remains; quantitation of T cell phenotypes may correlate with disease, but many times, the specific antigen is unknown, which makes it difficult to confirm exact

TABLE 7.2 Some Clinical Diagnostic Tests for Autoimmune Disorders

General tests for autoimmunity

C-reactive protein
Autoantibody titers (anti-DNA, antinuclear components)
Presence of specific HLA antigenic alleles
Increased erythrocyte sedimentation rate
Presence of rheumatoid factor

Specific antibody tests for autoimmunity	Indicated disorder[a]
Antithyroglobulin antibody	Autoimmune thyroid disease Hashimoto's thyroiditis
Antithyroglobulin stimulating hormone receptor antibody	Graves' disease
Antigliadin, antitransglutaminase, or antienodomysium antibodies	Celiac disease
Antimitochondrial antibody	Autoimmune liver disorder
Anti-islet cell antibody	Insulin-dependent diabetes mellitus (type I)
Antimyelin basic protein antibody	Multiple sclerosis
Anti-acetylcholine receptor antibody	Myasthenia gravis
Antidesmoglein 1 or 3 antibodies	Pemphigus (bullous dermatosis)
Antinuclear antibodies	Systemic lupus erythematosus Rheumatoid arthritis Scleroderma Sjögren's syndrome
Anti-double-stranded DNA (dsDNA) antibody Antiphospholipid/cardiolipin antibody	Systemic lupus erythematosus
Anticyclic citrullinated peptide antibody Presence of rheumatoid factor	Rheumatoid arthritis

[a] *Overlapping diagnostic identifications for specific disorders.*

disorders by direct diagnostic testing. Combination of all these tests with histology and pathology allows a more accurate assignment of the specific disease state.

TARGETED THERAPEUTICS

General immune-suppressing agents are often used to limit inflammation and cellular activity during autoimmune episodes (a thorough discussion of immunosuppressive therapy will be provided in Chapter 11).

As our understanding of the pathogenesis of autoimmune disorders increases, so does the potential to utilize targeted therapeutics to mediate specific immune events related to clinical development of tissue damage. For example, TNF-α inhibitors are now becoming increasingly useful for the treatment of rheumatoid arthritis, ankylosing spondylitis, psoriasis, and inflammatory bowel diseases.

Likewise, antibody-based therapeutics that interfere with TNF signaling, or IL-1 and IL-6 signaling, are tremendously powerful tools to reduce inflammatory responses. Interleukin-17 (IL-17) inhibitors, as well as targets against its receptor, are now being used for the treatment of autoimmune diseases such as rheumatoid arthritis, psoriatic arthritis, and ankylosing spondyloarthritis. Antibodies also can be targeted to limit active B-cell populations; for instance, the anti-CD20 antibody (a prominent B-cell surface antigen) is a very useful tool for the treatment of rheumatoid arthritis. In a similar manner, antibody biologics that bind the cytotoxic T-lymphocyte antigen 4 (CTLA-4; CD152) have been shown to downregulate interleukin-2 (IL-2)–producing CD4+ T-helper populations. Other therapeutics target innate functions to limit inflammatory profiles; beta-interferon (IFN-β) is useful to treat multiple sclerosis, and inhibition of type I interferons may eventually be used to treat systemic lupus erythematosus.

SUMMARY

- Autoimmunity represents a failure of effective tolerance to self-antigens.
- The development of autoimmunity results from the failure to effectively eliminate self-reactive lymphocytes and contain those lymphocytes after they enter peripheral tissues.
- Genetic and environmental factors play a role in the etiology of autoimmune disease.
- Mechanisms of disease include autoantibodies that are directed against specific self-components, deposition of circulating antibody-antigen complexes, and deleterious responses caused by autoreactive T cells.

Immune Hypersensitivities

CHAPTER FOCUS

To investigate concepts associated with disorders that are classified as immune hypersensitivity. Underlying mechanisms will discuss the basis of responses that culminate in clinical symptoms, focused on the primary cellular and molecular components for the basis of pathology development. The discussion will include allergic (Type I), cytotoxic (Type II), immune complex (Type III), and delayed-type hypersensitive (DTH) responses (Type IV), concentrating on how a vigorous immune response contributes to tissue damage.

HYPERSENSITIVE DISORDERS

Immunological diseases can be grouped into two large categories of deficiency and dysfunction. As discussed previously in Chapter 6, immune deficiency disorders are the result of the absence (whether congenital or acquired) of one or more immune system elements. In contrast, disorders due to immune dysfunction happen when a particular subset of immune responses occur that are detrimental to the host. This response may be against a foreign antigen or self-antigen and is usually defined as inappropriate regulation of an effector response. This happens in the absence of protection against pathogenic organisms. Notwithstanding, the host is adversely affected. A healthy immune system occurs as a result of a balance between innate and adaptive immunity, cellular and humoral immunity, inflammatory and regulatory networks, and small biochemical mediators (cytokines). Investigators understood the nature of this imbalance as related to pathology, and in the late 1960s, researchers and clinicians classified these dysfunctional immune responses into categories called **hypersensitivity diseases**.

The term **hypersensitivity**, therefore, refers to a definable immune response that leads to deleterious host reactions rather than protection against disease. Hypersensitivities are a major cause of clinical disease. Although the mechanisms can be defined for each subclass, in reality, there is considerable overlap in the underlying causes that contribute to the hypersensitive responses and how they adversely affect tissues in the body. The hypersensitivity reactions fall into four classes based on their mechanisms and the ability to passively transfer response through antibodies or T lymphocytes. These responses include inappropriate antigenic response, excessive magnitude of response, prolonged duration of response, and innocent bystander effector reactions leading to tissue damage. The major mechanisms are detailed next.

TYPE I HYPERSENSITIVITY: IMMUNOGLOBULIN E (IgE)-MEDIATED IMMEDIATE HYPERSENSITIVITY

Type I hypersensitivity is due to the aberrant production and activity of IgE against normally nonpathogenic antigens (commonly called **allergens**) (Fig. 8.1). This **allergic hypersensitivity** is also called **immediate hypersensitivity** because of the speed of reaction development. Mast cells and basophils ordinarily have high-affinity IgE receptors that are constitutively filled with that immunoglobulin isotype. Antigenic exposure results in a cross-linking of cell-bound IgE with the allergen, followed by nearly immediate activation of those cells. This results in a quick release of both preformed (e.g., histamine, leukotrienes) and newly synthesized mediators (e.g., chemotaxins, cytokines). The activity of these mediators is responsible for the signs and symptoms of allergic diseases. Note that the IgE-associated responses differ from the mechanisms of anaphylatoxins (complement factors C3a, C4a, and C5a), which trigger mast-cell degranulation in the absence of IgE.

For any Type I reaction to occur, there must be a preexisting, specific IgE population for the allergen. By definition, a mature B-cell response to the antigen has already developed, in part with CD4+ T-helper-2 cytokines such as interleukin-4 (IL-4) and interleukin-13 (IL-13), which promote the generation of plasma cells that produce antibodies of the IgE isotype.

The prototype disorders for this hypersensitivity include allergic rhinitis and seasonal allergies, as well as allergic asthma. Typical allergens include pollens, fungal spores, common dust mites, and other household antigens. Immunotherapy for severely reacting individuals usually includes the diversion and development of immune responses other than those related to IgE synthesis. Pharmaceuticals have been developed to inhibit mast cell degranulation, alleviating many common symptoms. In severe cases where allergens are systemic, anaphylactic shock may

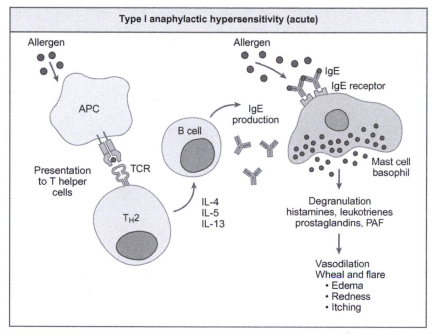

FIG. 8.1 Type I hypersensitivity (also called *immediate hypersensitivity*) is due to aberrant production and activity of IgE directed toward normally nonpathogenic antigens (commonly referred to as an *allergy*). The IgE binds to mast cells via high-affinity IgE receptors. Subsequent antigen exposure results in cross-linking of mast cell-bound IgE. Activated mast cells release preformed mediators which are responsible for allergy symptoms.

occur due to immediate vasoactive mediator activity. This is characterized by a sudden and sharp drop in blood pressure, urticaria (hives), and breathing difficulties caused by exposure to a foreign substance (such as bee venom, immediate drug reactivity, or food allergies). **Systemic allergic anaphylaxis** is life threatening; emergency treatment includes epinephrine injections used as a heart stimulant, vasoconstrictor, and bronchial relaxant.

TYPE II HYPERSENSITIVITY: ANTIBODY-MEDIATED CYTOTOXIC HYPERSENSITIVITY

Type II hypersensitivity is due to antibody reactivity against cell membrane-associated antigens that results in cytolysis (Fig. 8.2). The mechanism may involve complement (**cytotoxic antibody**) or effector lymphocytes that bind to target cell-associated antibody and effect cytolysis via a complement-independent pathway (**antibody-dependent cellular cytotoxicity, ADCC**). The result of the antibody response is cytolysis.

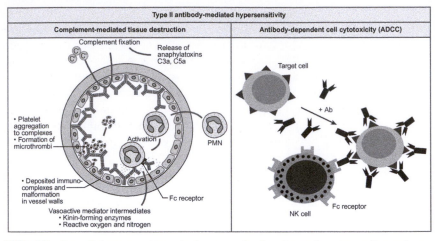

FIG. 8.2 Type II hypersensitivity is due to antibodies directed against cell membrane-associated antigens, which results in cytolysis. The mechanism may involve a complement (cytotoxic antibody) that binds to a target cell-associated antibody, subsequently effecting cytolysis *(depicted)* or via a complement-independent pathway of ADCC involving NK cells and targeted cell destruction *(not shown)*.

In complement-mediated Type II hypersensitivity, immunoglobulin G (IgG) isotype antibody recognition of cell surface epitopes leads to the assembly of the complement C5–C9 **membrane attack complex (MAC)** and subsequent lysis of the cell. This reaction is the underlying mechanism in multiple disease states, including those seen in autoimmune hemolytic anemia and Rh incompatibility leading to erythroblastosis fetalis. In rare cases, pharmaceutical agents, such as penicillin or chlorpromazine, can bind cells, forming a novel antigenic surface complex that provokes antibody production and Type II cytotoxic reactions.

A second mechanism for Type II reactions is characterized by ADCC induced by natural killer (NK) cells recognizing IgG attached to target cells bearing these antigens. The constant portion of the antibody (Fc region) is bound by Fc receptors on the NK cell, leading to perforin release and NK cell-mediated lysis. Neutrophils, eosinophils, and macrophages also may participate in ADCC, and it may be involved in the pathophysiology of certain virus-induced immunological diseases, such as those seen during active responses to retroviral infection.

Cytotoxic antibodies mediate many immunologically based hemolytic anemias, while ADCC may be involved in the pathophysiology of certain virus-induced immunological diseases. Prototype disorders include many autoimmune-related diseases that show evidence of tissue destruction. Goodpasture syndrome represents autoantibodies against basement membrane collagen type IV; deposition and accompanying complement activation lead to damage to both kidney and lung tissues.

Mediators of acute inflammation generated at the tissue form end-product MACs that cause cell lysis and death. The flared reaction takes a period time ranging from hours to a day, with chronic strain on the body if it continues unabated. Other disorders include idiopathic thrombocytopenic purpura (platelet destruction) and pemphigoid reactions resulting in skin blisters. And both myasthenia gravis (muscle weakness) and Graves' disease (hyperthyroidism) exhibit autoantibody-mediated cytotoxic events, although these disease processes are considerably more complex in nature and overlap with other hypersensitivity mechanisms.

TYPE III HYPERSENSITIVITY: IMMUNE COMPLEX-MEDIATED HYPERSENSITIVITY

Type III hypersensitivity results from soluble antigen-antibody immune complex deposition and subsequent events that activate complement to summon polymorphonuclear leukocytes (Fig. 8.3). The antigens

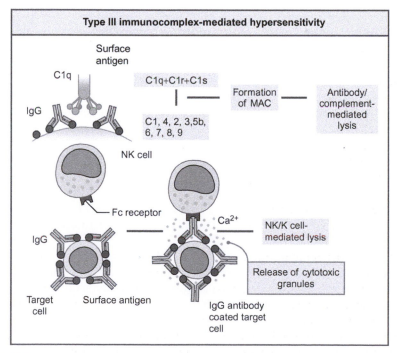

FIG. 8.3 Type III hypersensitivity results from soluble antigen-antibody immune complexes that activate complements. The antigens may be self or foreign. Complexes are deposited on the membrane surfaces of organs. The by-products of complement activation are chemotaxins for acute inflammatory cells.

may be self or foreign (i.e., microbial). Such complexes are deposited on the membrane surfaces of various organs (e.g., kidney, lung, synovium). The by-products of complement activation (C3a, C5a) are chemotaxins for acute inflammatory cells, resulting in infiltration by polymorphonuclear cells (PMNs). Lysosomal enzymes are released, which results in tissue injury. Platelet aggregation occurs, resulting in microthrombus formation in the vasculature. This type of hypersensitivity was classically characterized as the **Arthus reaction**, identified by a high degree of neutrophils and mast cell infiltrates, vasoactive amine release, erythema, and edema in response to intradermal injection of antigen.

These immune reactions result in Type III inflammatory injury, readily seen in diseases such as rheumatoid arthritis, systemic lupus erythematosus, and postinfectious arthritis. It is also evident during poststreptococcal glomerulonephritis, where damage severely affects kidney function.

An interesting example of Type III reactions is the condition referred to as **serum sickness**, which historically arose when antisera made in animals were repeatedly given to humans to neutralize toxins. Immune complexes would form in vivo after 1–3 days, when host B cells recognized circulating foreign (heterologous) antibodies, and specific antibodies were produced that targeted those foreign epitopes. Subsequent particulates deposited in vascular beds, leading to pathologies. Symptoms included anaphylactoid purpura, rash, fever, myalgia, and arthralgia. This type of reaction is still possible when intravenous immunoglobulin (IVIG) is administered therapeutically, although matching homologous human proteins limits reactivity. It is noteworthy that individuals can exhibit similar pathologies in reaction to drugs; complexes of antibodies directed toward chemotherapeutic agents (e.g., penicillin) may lead to immune complexes that deposit in vascular beds and result in vasculitis, destruction of endothelium, and edema.

TYPE IV HYPERSENSITIVITY: DELAYED-TYPE (CELL-MEDIATED) HYPERSENSITIVITY

Type IV hypersensitivity, also called **delayed-type hypersensitivity (DTH)**, involves T cell-antigen interactions that cause activation and cytokine secretion (Fig. 8.4). This type of hypersensitivity requires sensitized lymphocytes that respond 24–48 h after exposure to soluble antigens. DTH reactions may involve T-helper cells (CD4+) or cytotoxic T cells (CD8+ CTLs), rather than antibodies. Diseases such as tuberculosis, leprosy, and sarcoidosis, as well as contact dermatitis, are all clinical examples where tissue injury is primarily due to the vigorous immune response to released antigens, rather than damage due to the inciting pathogen itself. In these examples, sustained release of antigen and

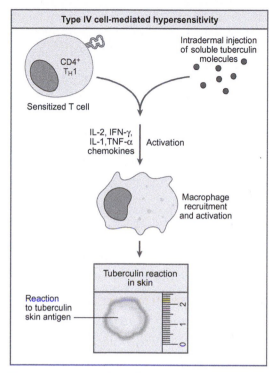

FIG. 8.4 Type IV hypersensitivity (also called *DTH*) involves macrophage-T cell-antigen interactions that cause activation, cytokine secretion, and potential granuloma formation. Shown here is the response to soluble tuberculin antigens. Although CD4+ T cells are depicted, they may give aid to cyotoxic CD8+ T cells to further the response. The tissue injury is primarily due to the vigorous immune response rather than the inciting antigen.

continued activation of sensitized T cells result in amplified tissue damage. Persistent infections may provoke excessive macrophage activation and granulomatous responses, leading to extended fibrosis and necrosis of tissue.

A classic DTH reaction is exemplified in the tuberculin skin test (i.e., the Mantoux reaction). An individual sensitized to tuberculosis through exposure or infection develops CD4+ lymphocytes that are specific to mycobacterial antigens. Intradermal injection of purified protein derivative (PPD) from mycobacteria results in the activation of sensitized CD4+ T cells. This is followed by the secretion of cytokines, which cause macrophage recruitment and activation. The final outcome is a localized reactivity manifested by erythema and induration. It is of interest that individuals who are HIV positive with low CD4+ cell counts will not mount significant DTH responses during the tuberculin skin test, providing further evidence of the importance of this hypersensitive reaction in protective immunity.

Type IV reactions also can be identified in pathologies resulting from viral infection. Here, CD8+ cells react to antigens presented via class I MHC molecules or to antigens associated with the cell surface. Cytotoxic cells recognize the presented antigen and lyse the infected target. Bystander killing may occur due to overaggressive responses, such as those seen during smallpox, measles, and herpes infections, as well as in contact dermatitis.

ALTERNATIVE HYPERSENSITIVITY CLASSIFICATIONS

Many investigators have revisited the four classifications discussed here, especially as the molecular knowledge of immunology has increased, and our understanding of the clinical manifestation of disease has grown as a result. A fifth classification encompasses pathology leading to the granulomatous response, in which the encapsulation and isolation of specific pathogens lead to tissue pathology. These events are driven by innate immunity or foreign-body responses leading to aggressive CD4+ Th1- or CD4+ Th2-type responses. The effect of the cytokines produced depends in large part upon the presentation of the particular persisting antigen.

SUMMARY

- Robust dysfunctional immune responses may often lead to tissue damage that is detrimental to the host.
- Allergic hypersensitivity (Type I) is an immediate hypersensitivity due to aberrant activity of IgE against normally nonpathogenic allergens, mediated through mast cell and basophil high-affinity IgE receptors.
- Cytotoxic hypersensitivity (Type II) results in tissue injury due to antigens recognized by IgG and IgM antibodies. Complement deposition triggers the formation of the membrane attach complex (MAC), causing lytic damage.
- Immune complex hypersensitivity (Type III) represents a disorder in which reactivity in vivo leads to the deposition of immune particulates. Complement by-products initiate chemotaxic influx of acute inflammatory cells, which release enzymes that result in tissue injury.
- DTH (Type IV) involves T-helper cells (CD4+) or cytotoxic T cells (CD8+ CTLs) rather than antibodies. This is a cell-mediated event directed at released antigens in a sensitized individual.

Vaccines and Immunotherapy

CHAPTER FOCUS

To emphasize immunological principles as they relate to vaccination. The goal is to develop a perspective about active and passive immunization via vaccination against infectious agents of multiple classes. The discussion will demonstrate how principles in immunology combine with biotechnology to advance the field of vaccinology. Information regarding immunotherapeutics is presented as a way to provide homeostasis of normal immune function.

PRINCIPLES OF VACCINATION

The definition of *immunity* is centered on protection against infectious disease. This may be conferred most readily by immune responses generated through immunization or previous infection. Edward Jenner, an English country doctor, observed that people infected with cowpox virus often developed less severe smallpox disease. Consequently, he inoculated a boy with cowpox virus obtained from hand sores on a milkmaid. Six weeks later, after the boy recovered from cowpox, he was reinoculated with virulent smallpox virus. The boy survived. Thus was born the process of **vaccination**. Since that time, similar methodologies have been successfully adopted for **immunization** against a multitude of diseases.

Although unknown at the time, Jenner's vaccination represented **cross-reactivity** of common antigens present on the cowpox virus with molecules present on the smallpox virus. Antibodies raised against the avirulent form also were able to neutralize virulent infection. It has been shown subsequently that the development of specific antibodies is a powerful tool to provide long-lasting immunologic protection against infectious agents. Indeed, it is now appreciated that a wide variety of responses are triggered via immunization; specific pathways can be targeted to elicit the arm of the immune response that is most critical for protection against distinct pathogens (Table 9.1). Major advances in vaccine design are taking place. Improvements in methodologies to produce nonvirulent antigenic substances for use as vaccine antigens will dictate future successes in the immunization arena. These include novel ways to manufacture toxoids and synthetic peptides, improvements in recombinant DNA technology to allow live, avirulent (nondisease-causing) viral and bacterial agents to express other pathogen genes, development of DNA-based vaccines, and new methods of conjugation to achieve superior immunogenicity for both polysaccharide and protein antigens.

BASIC CONCEPTS OF PROTECTIVE IMMUNIZATION

The objective of immunization is to generate high levels of memory cells using vaccination methods (Fig. 9.1). The term **primary immune response** refers to a lymphocyte activation event following first recognition of foreign material, following which a memory response is generated. Immunological memory represents a pool of circulating, long-lived cells that remain present and available for action long after the initial response activities wane. If the antigen is reencountered at a later time, a **secondary immune response** occurs, in which memory cells are engaged and activated. This secondary response is faster, more focused, and more effective than the original encounter.

The development of disease is a complex equation that gives rapidity of response as a critical component related to disease outcomes. The

TABLE 9.1   Vaccine Classes and Their Targets

Type of vaccine	Components	Examples
Live, attenuated	Viral or bacterial organism with reduced pathogenicity	Oral polio, varicella, measles-mumps-rubella (MMR), bacillus Calmette-Guerin (BCG)
Killed-inactivated	Whole killed organism	Inactivated polio, typhoid
Subunit Recombinant subunit	Inactivated or modified toxins, purified components Gene-derived proteins produced in another organism	Diphtheria, tetanus, influenza Hepatitis B toxoid, human papillomavirus (HPV)
Conjugate/polyvalent	Combined components isolated or genetically modified from multiple strains	*Haemophilus influenzae* type B, *Streptococcus pneumonia, Neisseria meningitidis*

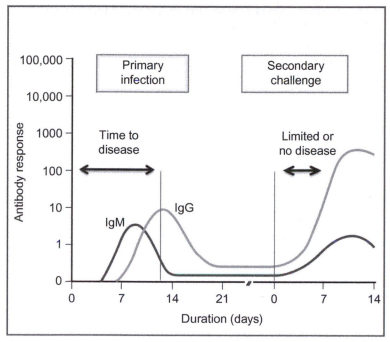

FIG. 9.1 Primary and secondary immune responses. A primary infection generally leads to a stronger and more rapid elicitation upon secondary exposure. However, if the incubation time is short during primary exposure, the development of secondary responses is impaired. Vaccination permits preexisting responses to dominate, which is critical during short infection incubation periods.

incubation period during the establishment of infection is important because it dictates how much time is available to mount an immune response prior to disease initiation. A long pathogen incubation time results in an extended period where immune events can mature. This naturally leads to induction of a relatively stronger immune response, with significant development of immunological memory. In this case, secondary exposure results in protection to disease due to preestablished memory responses. However, when there is only a limited incubation period existing a short time prior to disease induction, there is only a limited time for inducting immune responses. When this happens, secondary exposure does not necessarily elicit protection against disease. Vaccines are especially useful in this example, as there is a shorter window of opportunity for fighting pathogens with them. Induction of protection can be achieved with a vaccine, provided that one is able to sustain high, long-term, reactive antibody **titers**. This is clinically achievable by giving several immunizations in a shorter time frame to raise sustained titers of specific antibody responses.

TYPES OF IMMUNIZATIONS

Immunizations may be active or passive. Active immunization is a result of direct exposure to an antigen, which allows the host to generate protective immunity. The objective is to provide long-lasting immunity against future exposures. **Active immunity** may be acquired naturally, through infection (and subsequent recovery), or artificially, through vaccination. **Passive immunity** also provides protection for the host, but it is conceived by the administration of humoral and/or cellular factors that provide immunity for the host. In this case, the host does not actively generate a protective response. The objective of passive immunity, therefore, is to provide immediate but temporary protection against an imminent or ongoing threat.

AGE AND TIMING OF IMMUNIZATIONS

Vaccines are given to prevent life-threatening infections. It is critical, therefore, to relate factors such as patient age, demographics, geographical location, and pathogen incidence to the vaccines being administered (Table 9.2). For example, neonate or pediatric vaccines typically target pathogens that rapidly outpace an infant's ability to respond effectively. The newborn is naturally delayed in the development of immune responses, especially in the production of immunoglobulin isotypes (Fig. 9.2). Newborns are immunocompetent but immune immature; fetuses make immunoglobulin M (IgM), but not immunoglobulin G (IgG) until birth. Maternal IgG provides protection against bacterial agents through the first months of life. While the total amount of all immunoglobulin IgG in newborn serum is at a level (mg/ml) close to that of a normal adult, almost all of it is of maternal origin. The half-life of IgG is 2–3 weeks; only 10% of maternal antibodies remain by 4 months of age, and only 3% by 6 months. Fortunately, at 2–3 weeks postpartum, supplemental antibodies of the immunoglobulin A (IgA), IgM, and IgG isotypes are delivered through colostrum and breast milk.

Children under 2 years of age remain immunologically disadvantaged, with limited ability to produce antibodies other than those of the IgM isotype to bacterial capsular polysaccharides (T-independent antigens). Vaccines are designed to work with the newborn's developing immune system to elicit opsonizing antibodies of the IgG isotype. One trick is to link a polysaccharide molecule, or a hapten, to a carrier protein chemically to enlist a strong T-helper-cell response and induce accompanying antibody isotype-switching.

TABLE 9.2 US Schedule for Active Immunization of Children and Adults

Vaccine (birth to 18 years old)	Age (dosing dates for vaccination)
Hepatitis B	Birth, 1–2 months, 6–18 months
Rotavirus	2 months, 4 months
Diphtheria, tetanus, acellular pertussis (DTP)	2 months, 4 months, 6 months, 15 months, additional dose after year 4
H. influenzae type b (Hib); pneumococcal conjugate (PCV13)	2 months, 4 months, 6 months, 15 months, additional dose after year 1
Poliovirus	2 months, 4 months, 6–18 months, and after year 4
Influenza	Annual vaccination after month 6
MMR; varicella	1 year, additional dose after year 4
Hepatitis A	1 year, additional dose before year 2
Influenza (childhood)	Annual vaccination after month 6
HPV	3 doses in early teenage years
Tetanus, diphtheria, pertussis (Tdap)	1 dose annually, begin year 7, boost in year 11 and each decade after
Meningococcal	11 years, boost in year 16
Vaccine (19 years old plus)	Age (dosing dates for vaccination)
Influenza (adult)	1 dose annually, begin year 19
HPV	2–3 doses over lifetime, depending on age at series initiation
Tetanus, diphtheria, pertussis (Tdap)	1 dose annually, boost each decade
Varicella	2 doses recommended over lifetime
HPV	Boosting of individuals of high risk or immunocompromised
Pneumococcal conjugate (PCV13 or PPSV23)	Boosting of individuals of high risk, plus after year 65
Meningococcal	2 doses recommended over lifetime
Hepatitis A, hepatitis B	2 doses recommended over lifetime
H. influenzae type b (Hib)	1 or more doses dependent upon indication
Zoster	50–60 years old, or older

An expanded recommended immunization schedule maintained by the Centers for Disease Control and Prevention (CDC) may be found at http://www.cdc.gov/vaccines/schedules/index.html.

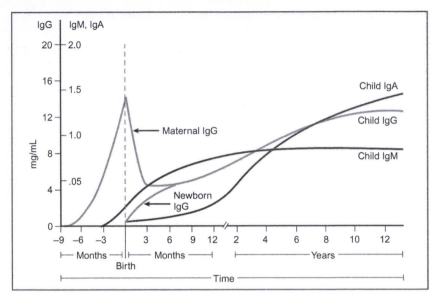

FIG. 9.2 Changes in relative antibody titers and isotypes in the newborn.

On the other end of the age spectrum, the elderly (i.e., > 60 years of age) exhibit a reduced capacity to mount primary responses to most antigens. Extreme age is a determinant for immune regulation; immune senescence occurs, in which the majority of memory responses remain available but poor primary (naive) response results in increased susceptibility to organisms and strains never before encountered. It is noteworthy that the immune-senescent individual retains a strong response to bacterial polysaccharides. The goal with elderly individuals, therefore, is to induce high levels of specific responses. It may be necessary to repeat vaccinations at more frequent intervals (years, rather than decades) to maintain a high functional response level.

In healthy individuals, multiple doses may be required to induce immunoglobulin isotype switching and to attain high levels of antibody titer sufficient for long-lasting protection. The critical factor is knowledge of which arm of the immune response is required to optimize protective responses against any particular infectious agent or pathogenic factor. Induction of B-cell responses for antibody production is a successful method of toxin or viral neutralization. Antibodies are also tremendously effective for opsonizing bacteria to prepare them for phagocytosis, as well as for targeted pathogen killing when combined with complement components.

Vaccines that drive T-cell development are effective against both intracellular and extracellular agents and to help monocytes to stimulate inflammatory responses that are critical to destroying invading pathogens.

Cytotoxic lymphocytes are necessary for the targeted destruction of intracellular pathogens, such as viral agents, in which a lethal hit delivered to infected target cells is required to limit the spread of infection. Adjuvants, discussed later in this chapter, are essential for directing the immune response in a way that will benefit protective outcomes.

Vaccines are especially effective for use in selected populations. These include military personnel, as well as travelers to high-risk areas, who may be exposed to pathogens not typically encountered in local environments. Vaccines will be an especially important tool to control outbreaks (and future outbreaks) of rare illnesses, and this is especially urgently needed for the Ebola virus, the Zika virus, and the Middle East Respiratory Syndrome (MERS)- and Severe Acute Respiratory Syndrome (SARS)-associated coronavirus agents.

Similarly, veterinarians and animal handlers should consider vaccination against pathogens found in the workplace. College students should be vaccinated against sexually transmitted diseases and against agents that are easily spread in high-contact areas. Obviously, health care workers and physicians should be vaccinated against pathogens transferred through accidental blood product exposure. Also, although not completely obvious, it is recommended that those in high-risk lifestyles [e.g., high human immunodeficiency virus (HIV) demographics or intravenous (IV) drug users], as well as those that are immunocompromised, should undertake regular reviews of vaccines for prophylactic protection. Finally, while less common, therapeutic vaccines may be used to treat infected individuals with slow-growing pathogens. These include vaccines that may be used as a targeted therapeutic to treat diseases such as rabies and hepatitis.

A short but frank discussion of relative vaccine safety is warranted. Vaccines are a safe and proven mechanism to induce strong protective immunity. They can elicit minor temporary inflammation following administration; however, those effects are limited in duration and quickly dissipate (2–3 days) postinjection. In some cases, elevated levels of hypersensitivity can occur, and there are rare cases of arthritis and arthralgia in individuals prone to these reactions. The US Food and Drug Administration (FDA) set strict rules for the removal of toxic products during formulation of vaccines. The current public reports of deaths in infants and the elicitation of neurological impairment due to vaccination are false. In addition, there is no proven link of infant vaccines to developmental disorders, such as those causing autism; data to back up those claims were falsified and the report has since been retracted.

While increased hygiene is protective, vaccines are critical to keep disease at bay. **Herd immunity** is a social concept relating that vaccination of a significant portion of a population (or "herd") gives a measure of protection for individuals who have not developed immunity; vaccines protect us as well as those around us due to the subsequent limitation of infection spread. Multiple vaccinations work; the child's immune system is robust and cannot be overwhelmed by multiple immunizations. Indeed, a child is exposed to more antigens in their daily routine than through all the vaccines combined that they receive during childhood. Care should be taken with live vaccines, even if the organisms are attenuated. They should not readily be given to immunocompromised patients or to those with severe immune disorders. Likewise, patients undergoing concurrent immunosuppressive therapy, or even pregnant women, should avoid being vaccinated with live organisms.

Immunologic adjuvants are excipient components added to vaccines to potentiate immune responses. In essence, they function to direct antigenic responses toward achieving the desired immune outcome and allow the vaccine to be a more effective prophylactic candidate. In general, adjuvants are capable of assisting in the generation of immunity, through stabilization of the antigen delivered and direct triggering of cellular responses.

The only adjuvants approved for use in the United States are the mineral salts (**aluminum salts; alum)** and an oil-based emulsion capable of stabilizing functional delivery of the antigen. However, many novel molecules are under active research for inclusion as approved adjuvants in vaccine preparations. They include plant saponins, cytokines, bacterial cell wall products, particulates, viral-like particles, and nucleic acid motifs. Some function by stimulating antigen-presenting cells (APCs) through interaction with pathogen-associated molecular patterns (PAMPs) or Toll-like receptors (TLRs). Others function to stabilize the antigen for targeted presentation by APCs to T lymphocytes.

PASSIVE IMMUNIZATION

An example of passive immunization was mentioned previously when describing how newborns are protected with maternal IgG. As a therapeutic class, passive immunization is an extremely useful clinical tool. Immunoglobulins are routinely given to patients to prevent or treat disease, and they are especially potent in immune-deficient individuals. Pooled antibodies from immune donors can be given intravenously. These **intravenous immunoglobulins (IVIG)** used to treat primary immune deficiencies can restore regular immune function, although continued administration is required every 3–4 weeks due to the half-life of the IgG isotype given.

Historically, antibodies raised in animals also could be administered to humans to treat infections. For example, antibodies against tetanus toxoid isolated from serum of immunized horses could be given to neutralize the tetanus toxin. Unfortunately, the heterologous antisera, being nonhuman in nature, were recognized as foreign when given more than once. This led to hypersensitive responses.

Fortunately, science has evolved the technology to produce monoclonal antibodies in the laboratory that are homologous in their physical amino acid structure (antibodies of the same species), which limits cross-reactivity to heterologous epitopes. The FDA now has a panel of approved antibody-based products for passive immunization and targeted immunotherapy, many of which have been "humanized" (via molecular engineering) to include human-constant regions while retaining the antigenic specificity targeted by the original immunoglobulin. This allows the monoclonal to be used with little or no chance of foreign reactivity developing.

THERAPEUTIC USES OF IMMUNOGLOBULINS

The ability to utilize passive antibodies as therapeutic agents has exploded in the last decade. In addition to IVIG, monoclonal antibodies have been successfully used to treat multiple autoimmune disorders. Specifically, antibodies that either neutralize or inhibit the binding of tumor necrosis factor (TNF) to its receptors are important therapeutic tools to fight the manifestation of pathology during autoimmune responses. Likewise, monoclonals can play a targeted role in the fight against cancers of both hematologic and solid tumor forms. These directly target B-cell malignancies, breast cancer, chronic myelogenous leukemia (CML), and chronic lymphocytic leukemia (CLL), to name just a few.

Another example for antibody use as a therapeutic agent is of clinical importance in a specific complication of red blood cell (RBC) surface antigen incompatibility between mother and fetus. Rh antigens, also called *Rhesus antigens*, are transmembrane proteins expressed at the surface of erythrocytes. They appear to be used for the transport of carbon dioxide (CO_2), ammonia, or both across the plasma membrane. RBCs that are Rh positive express a specific antigen designated the D type (RhD antigen). About 15% of the population have no RhD antigens and thus are Rh negative. An Rh-negative mother who carries a Rh-positive fetus runs the risk of producing immune antibodies to the Rh antigens on the fetal RBC. The exposure during primary pregnancy is minimized. However, the mother may generate Rh antibodies after birth if the mother comes into contact with fetal blood cells during placenta rupture. Some fetal RBCs enter the mother's bloodstream, thus allowing the production of maternal-derived anti-Rh antibodies.

Upon subsequent pregnancies, the next Rh-positive fetus also will be at risk because the mother will retain a low level of circulating antibodies against the Rh antigen. Destruction of fetal erythrocytes will occur via the passive immune transfer of maternal antibodies to the fetus, resulting in erythroblastosis fetalis (a hemolytic disease of the newborn). It is of great clinical importance to identify Rh-mismatched mothers and fetuses. If there is such a mismatch, the mother is clinically treated with anti-Rh antibodies [Rh immune globulin (RhIG) or Rhogam], which react with the fetal RBC. The ensuing antibody-antigen complexes are removed prior to maternal recognition of foreign Rh antigens.

Monoclonal antibodies are now being used regularly to treat diseases with immunological etiology. The essence of this technology (mentioned previously in this chapter and in Chapter 7) takes advantage of the exquisite binding capabilities of antibody-variable domains to recognize specific biological targets. These targets may be receptors on cells, or they may be inflammatory mediators released during disease. These biologically-based therapeutic agents have the advantage of being pro-duced in the laboratory under quality-controlled conditions. Recent advances in this field have included the extension of these therapeutics for use in nonimmune-related disorders.

OTHER WAYS OF MODIFYING IMMUNITY

Finally, it should be noted that the removal of pathogenic antibodies and other immune factors is an important immunotherapeutic option. Another example again relates to a child born to a mother suffering specific autoimmunity. Maternal antibodies are passed to the child, and the child exhibits disease symptoms that mimick those in the mother.

A specific case can be seen with a mother who has active Graves' disease; maternal antibodies passed to the newborn initiate relatively high activation of the thyroid gland, leading to clinical symptoms of hyperthy-roidism. Successful treatment can be accomplished by plasmapheresis, in which reactive maternal antibodies are filtered from the serum of the newborn; elimination of reactive antibodies eliminates clinical presenta-tion in the child.

SUMMARY

- Vaccines are a safe and proven mechanism to induce strong immunity. Immunization has had a tremendous impact on the quality and longevity of human life by eliminating devastating pathogens.
- Vaccines vary to accommodate the targeted immune recognition of pathogens in advance of encountering the infectious agent. Strategies incorporate the physical basis and nature of the antigen with stabilizing delivery vehicles. Adjuvants can be added to promote directed immune function.

- The nature of the antigen used in immunization is critical to eliciting the subsequent response. Technological advances in molecular production allow a broad range of antigens for use in vaccination.
- Passive administration of immunoglobulins or immune factors allows short-term protection in the absence of preexisting immune responses.

Cancer Immunology

CHAPTER FOCUS

To investigate natural (effective) responses to tumor development and formulate how immune components function to eliminate potentially dangerous precancerous events. This will be followed by a discussion of challenges faced when protective responses fail and tumors develop. Categories of tumor antigens will be described, related to their impact on self-recognition. A review of effector mechanisms in tumor immunity will be given, and there will be information on cancers of the immune system when the protective cells themselves become the cause of tumorigenic activity. Finally, aspects of immunodiagnosis and immunoprophylaxis will be addressed.

UNDERSTANDING IMMUNE DEFENSES AGAINST CANCERS

The field of cancer immunology investigates interactions between the immune system and tumors or malignancies. At the heart of this is the specific recognition of cancerous cells and the antigens they express. In essence, this is the concept of natural **immunosurveillance**, proposed in 1957 by F.M. Burnet and L. Thomas. They suggested that lymphocytes act as sentinels to recognize and eliminate continuously arising and nascent transformed cells. A stepwise immunosurveillance process ensues, beginning with the recognition and elimination of dysregulated cells; both innate and adaptive immune cells recognize and kill tumors at early stages during development. A balanced state of equilibrium can be assumed, representing the control of cellular and tissue growth. In time, there is the potential for cellular escape from naturally protective responses, resulting in tumor pathogenesis.

TUMOR ANTIGENS

At the heart of immunosurveillance is the concept that cancerous cells express tumor antigens that can be recognized by various immune cell phenotypes. Tumor antigens fall into general categories, which basically include many normal gene products that are turned on at inappropriate times or appear on cell types that normally do not express those particular proteins. Examples include a group of antigens produced in adult tissue that are normally present only during developmental stages. These are **oncogenes**, or **oncofetal antigens**, which are of embryonic origin. Additionally, gene products may be expressed that function to inhibit growth suppression, allowing uncontrolled expansion of that particular cell phenotype.

The mechanisms in which mutant cellular gene products arise vary. A somatic mutation or point mutation may occur or a genetic rearrangement may happen. In either case, a change in protein function takes place, usually leading to alterations in normal cell function or protein structure. This can happen during cell cycle and replication, through improper DNA repair mechanisms, or by the activity of carcinogens. However, the result is typically unrestricted growth in a nonregulated manner.

Viral gene products also may trigger **oncogenesis**. Indeed, retroviral-encoded oncogenes may subvert normal cellular control mechanisms, turning the cell host into a vehicle for continued viral expansion. Along the way, the infected cells become cancerous. In addition, viral products can influence cellular expansion, directly via the introduction of point mutations or targeted host gene amplification, or by chromosomal translocation events that generate new gene products with untold consequences.

For simplicity, any overexpressed, mutated, dysregulated, or rearranged gene product expressed by a cancerous cell is considered a tumor antigen (sometimes referred to as a **tumor-specific transplantation antigen**). These tumor antigens play a role in differentiating tumors as being immunologically different from healthy tissue and allow the immune system to recognize cancerous cells as nonself. A **proto-oncogene** is a normal gene that undergoes mutation to allow increased expression. It may regulate other genes through functional activation or suppression. Activation of cellular proto-oncogenes in human cancers most often function in a way that affects cell growth. What is critical to keep in mind is that the protein is primarily a self-protein, albeit one that is expressed with a mutation or minor change in antigenic structure. Thus, these antigens are typically very difficult for detection by typical lymphocytic activation processes that deal with foreign antigenic recognition.

A special class of tumor antigens includes those proteins that are carcinogen-induced. **Carcinogens** can induce mutations in normal genes that were previously silent, giving rise to an array of different gene products. By definition, oncogenic and viral-induced mutations are similar between individuals; however, carcinogen-induced mutations appear antigenically unique. As such, there is very little or no cross-reactivity in these tumor antigens between individuals due to the random mutations induced by the chemical or physical carcinogenic substance.

EFFECTOR MECHANISMS IN TUMOR IMMUNITY

Both innate and adaptive immunities play defined roles in the fight to control tumor expansion (Table 10.1). The nature of the cancer is a significant factor in how the immune system functions to combat these rapidly expanding dysregulated populations. Similar to what was discussed in Chapter 5 in the fight against infectious agents, the location of the tumor or cancerous cell is critical to understanding immune mechanisms available to mount, direct, and complete a functional immune response.

The same is true for cancerous cells. Specifically, dispersed cells are far easier to target compared to solid tissue masses with little to no direct blood flow or common lymphatic drainage. But even solid tumors can be infiltrated by a broad range of immune cell types, in a productive manner, to limit tumorigenesis.

NATURAL KILLER (NK) CELLS AND INNATE RESPONSE TO TUMOR CELLS

Innate immune cells play a critical role in the antitumorigenic process. While natural killer (NK) cells, natural killer T (NKT) cells, and gamma-delta ($\gamma\delta$) T cells all have tumoricidal functions, the NK cells are most important for surveillance functions. NK cells are large, granular

TABLE 10.1   Immune Effector Mechanisms to Fight Tumor Cells

Effector mechanism	Activity
NK cells	Cytolysis, apoptosis, ADCC Lytic activity on tumors expressing low-MHC molecules
CD8+ cytotoxic T cells	Cytolysis, apoptosis Reject viral or chemical-induced tumors
CD4+ helper T cells	Help to CTLs Cytokine support for other effectors
B cells	Antibodies and complement-mediated lysis Antibodies for contribution to ADCC by NK cells
Macrophages/neutrophils	Presentation of altered self-antigens Cytokine production to support adaptive effector cells

lymphocytes that nonspecifically kill tumor cells. They express CD16 and CD56 and share many surface molecules with T lymphocytes but do not express specific antigen receptors. Essentially, they are non-T and non-B lymphocytes that lack surface CD3, CD4, CD8, and CD19. NK cells mediate the lysis of target cells by the release of lytic granules and perforin-induced pore formation. In this manner, they function like CD8+ CTLs, but the NK cells are able to kill "self" in the absence of antigenic recognition. A likely method involves "recognizing" the absence of histocompatability molecules on the target cell—a common occurrence in cancerous cells. NK cells also may kill target cells using Fas/Fas-ligand (FasL)-mediated apoptotic pathways. Note that killing by NK cells is enhanced by cytokines present in an inflammatory environment, such as alpha interferon (IFN-α), beta interferon (IFN-β), and interleukin-12 (IL-12). Further activation can occur in the presence of activated T cells. Activated NK cells produce interleukin-2 (IL-2), IFN-α, gamma interferon (IFN-γ), and tumor necrosis factor alpha (TNF-α).

Another mechanism for NK-cell activation utilizes a receptor for the constant portion of IgG; antibodies that recognize tumor targets are captured by the NK Fc receptor (CD16), which triggers the killing of target cells using an **antibody-dependent cellular cytotoxic (ADCC)** mechanism (Fig. 10.1).

ADAPTIVE RESPONSE TO TUMOR CELLS

Adaptive immunity plays a strong role in the elimination of tumor targets, especially when innate surveillance mechanisms are delayed. These mechanisms employ the specific antigen receptors on T cells, of both the CD4+ helper subset and the CD8+ cytotoxic varieties. Current thinking suggests that a T-helper type 1 effector response is required for effective antitumor activity. CD4+ T cells recognize antigens presented by MHC class II molecules on an antigen-presenting cell (APC), dendritic cell, or macrophage, following which they provide help through cytokines such as IL-2 and IFN-γ to activate CD8+ T cells.

It is likely that macrophages play a critical role in detecting cancer cell debris by phagocytosing cellular membranes followed by localized draining via lymphatics to regional lymph nodes. Helper T cells mature and provide productive assistance to CD8+ cells, which in turn recognize altered tumor transplantation antigens presented on target cells in the context of MHC class I histocompatability molecules. Cytotoxic lymphocytes function to kill their targets using perforin, granzymes, and destructive cytokines (TNF-β, IFN-γ). Recent evidence also suggests a role for Fas and FasL in the specific lymphocytic process to control tumor expansion.

Realize that T cells have natural **immune checkpoints,** in which surface molecules interact with local cells in a way that is designed to regulate

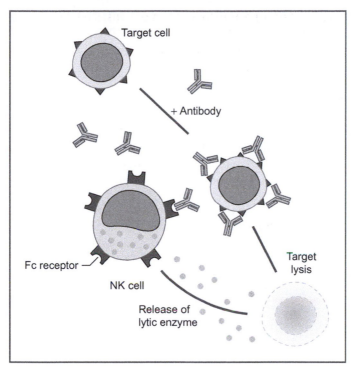

FIG. 10.1 ADCC is a phenomenon in which target cells coated with antibodies are destroyed by specialized killer cells, such as NK cells. The killing cells express receptors for the Fc portion of antibodies, which recognize tumor antigens, resulting in the release of lytic enzymes at the site of contact. Target cell killing also may involve perforin-mediated membrane damage.

immune activation; these molecules play key roles in preventing reactivity to self. Examples of some of these molecules include programmed cell death protein 1 (PD-1), cytotoxic T-lymphocyte-associated protein 4 (CTLA-4), and lymphocyte-activation gene 3 (LAG-3). By definition, tumors are self, albeit with changes that allow them to grow in an unchecked manner. This means that the immune checkpoints must be overcome to allow lymphocytes to perceive and eliminate tumors that arise in our bodies.

Antibodies play an adaptive role toward tumors. The cell-dependent mechanism involving antibodies and NK cells was mentioned previously. However, it is also critical to know that antibody and complement are other powerful tools to limit tumor cell expansion. As with all manners of antibody recognition, this depends upon the altered surface antigenic structures present on cancer cells—the greater the antigenic modification, the more likely that antibody-based mechanisms will be successful at limiting tumorigenesis.

ESCAPE MECHANISMS OF TUMOR ELIMINATION

The survival pressure for the expanding tumor cell population is high, and tumor-related "escape" from immune surveillance is common. Multiple mechanisms are the result of changes to the tumor itself, while other escape methods rely on alterations in host immune function. Regarding the tumor population, there are many instances where cancer cells demonstrate modulation of the expression of key target molecules, thus contributing to immune avoidance. This includes reduction of MHC molecule expression, change to antigenic nature of the tumor-associated surface antigens, and shedding of tumor antigens so that the cell is no longer recognized as an altered self cell.

Often, a deficiency in the host immune response allows the escape of tumor cells and resistance to attack. It is quite common for the tumor microenvironment to become infiltrated by immune suppressing cells, including T regulatory cells (Tregs) that downregulate functional responses. Additional cell phenotypes can be involved in the escape phenomenon as well. Myeloid-derived suppressor cells (MDSCs) represent a heterogeneous population that comprises myeloid progenitor cells, immature macrophages, immature granulocytes, and immature dendritic cells. Some of these phenotypes are quite adept at modulating tumor-associated macrophages, skewing their ability to present tumor antigens to incoming helper and cytotoxic T lymphocytes (CTLs). Finally, it should be noted that influences that affect systemic immunity in the host, such as infection or general immunodeficiency, also may contribute to the impairment of natural immunosurveillance function.

TUMORS OF THE IMMUNE SYSTEM

Problems also arise when the protective cells themselves become the cause of tumorigenic activity. Indeed, immune system cells are not actually immune from the dysregulated responses discussed previously. Although varied, the cancers generally fall into two classes: myeloid-based or lymphoid-based **leukemias**, representing cancers of the white blood cell (WBC) leukocytes. A partial list of common leukemias is discussed next.

Multiple myeloma is a cancer of plasma cells, a type of activated B-cell phenotypic for producing antibodies. In multiple myeloma, abnormal plasma cells will characteristically accumulate in the bone marrow, interfere with the production of normal blood cells, and damage parenchymal tissue beds. Most cases feature a diagnostic overproduction of a single type of immunoglobulin protein readily found in circulation (called an *M-protein*). **Burkitt's lymphoma** is another example of a B-cell cancer in which tumor cells from the patient possess identical immunoglobulin heavy-chain gene rearrangements. **B-cell chronic lymphocytic leukemia,** also known as **chronic lymphoid leukemia** (or CLL), is a common type of small lymphocytic lymphoma that can be distinguished based on maturity of the immunoglobulin variable-region heavy-chain (IgVH) gene mutation status.

Hodgkin's lymphoma is a defined type of lymphoma also originating from WBC leukocytes. It is characterized by spreading from one lymph node group to another. Microscopic examination reveals a characteristic histology, with the presence of multinucleated Reed–Sternberg cells. Various pathologic subtypes are possible, based on the cell morphology and composition of infiltrate seen within lymph nodes. Hodgkin's lymphomas may be characterized as distinct from **non-Hodgkin lymphomas (NHL),** which represent a diverse group of blood cancers that expand the definition of cancerous lymphomas. Types of NHL vary significantly, depending on the underlying cause (e.g., viral, genetic, iatrogenic). T-cell lymphomas generally fall into this latter category, usually associated with viral etiology.

Chronic myelogenous leukemia (CML) represents another type of leukocyte cancer, characterized by unregulated growth of predominantly myeloid cells in the bone marrow and subsequent accumulation of these cells in the blood. CML may be considered a clonal bone-marrow stem cell disorder. This cancer represents a myeloproliferative disease with the expansion of mature granulocytes of neutrophil, eosinophil, and/or basophil origin.

CML is associated with a characteristic chromosomal translocation called the *Philadelphia chromosome.* A chromosomal fusion event occurs between the Abelson (*Abl*) tyrosine kinase gene at chromosome 9 and the break point cluster (*Bcr*) gene at chromosome 22. A chimeric oncogene is created (called *Bcr-Abl*), with a resulting product implicated in disease pathogenesis. **Chronic neutrophilic leukemia (CNL)** is another chronic myeloproliferative neoplasm, albeit much less common. This leukemia represents myeloid hyperplasia in bone marrow, although there is an absence of the chromosome rearrangement found in CML.

IMMUNODIAGNOSIS AND IMMUNOTHERAPY

The use of antibodies targeted toward specific antigens is a powerful tool to histologically detect and identify tumors within tissue. Immunohistochemistry is a laboratory technique in which tissue sections can be screened for the presence of tumor antigens, allowing diagnostic identification of cancer cell phenotypes on surgical specimens. The ability to detect and classify the type of cancer accurately has great potential for subsequent therapeutic intervention. Likewise, the detection of specific tumor antigens on phenotypic cells allows accurate prediction of tumor aggressiveness, as well as its potential for responsiveness to therapeutic intervention.

Significant advances in our understanding of tumor immunology have led to a multitude of therapies based on the manipulation of the immune system (Fig. 10.2). Specifically, understanding which arm of the immune response is responsible for the rejection and destruction of tumors has led to the development of incredibly powerful clinical tools. For example, BCG immunotherapy is commonly used to treat early-stage bladder cancer; stimulation of innate responses allows local environments to respond in a positive manner to limit tumor growth. Similarly, Imiquimod, a topical therapeutic that supplements local production of IFN-γ, can be advantageous when accompanying radiation therapy or chemotherapy. Future therapeutics will stimulate innate responses, as exemplified in animal studies where the addition of nucleotide analogs stimulates Toll-like receptors (TLRs) such

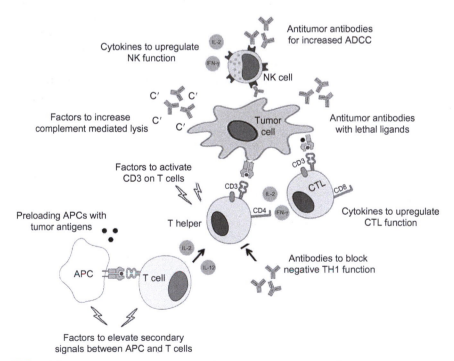

FIG. 10.2   Therapeutic interventions to boost anticancer immune function.

as TLR7 to augment cellular responses. Another example is the use of monoclonal antibodies, which are quite effective to target and destroy tumors.

Finally, a novel therapeutic advance has allowed the expansion of T cells to treat cancer. In this scenario, the patient's own T cells are used in a directed attack on tumor antigens. The cells are removed from the patient, reengineered in vitro in tissue culture, expanded, and then placed back within the patient. This engineering changes the T-cell antigen receptor in a way that forms a chimeric-binding receptor to recognize and target the cancerous cells directly. The technology, called **CAR T-cell therapy**, is aptly named after its use of the chimeric antigen receptor (CAR), and it is an excellent example of personalized medicine.

SUMMARY

- Tumor cells differ from normal counterparts by indefinite proliferation and changes in growth regulation.
- Normal cells can be transformed by chemical and physical carcinogens or by transforming viruses. They typically express tumor-specific antigens on their cell surface.
- Innate immune responses to tumors include NK-cell killing, ADCC, and macrophage-mediated cell killing. Adaptive immunity requires activation using specific receptors of both CD4+ helper and CD8+ cytotoxic lymphocytes. The CTL-mediated cell lysis is a particularly powerful and specific mechanism to control tumor growth.
- Some tumors cells utilize immune response-evading mechanisms.
- Cancer immune therapy includes monoclonal antibodies, antibodies coupled with toxins, chemotherapeutic agents, and radioactive elements.

Transplantation Immunology

CHAPTER FOCUS

To examine immune regulation of transfer, or grafting, of tissues from one person to another. Transplanted organs have the potential to be rejected by the recipient's immune system unless the person is either tolerant or immunosuppressed. Concepts associated with mechanisms underlying the immunobiology of transplantation will be discussed. The goals are to present genetic relationships between individuals that are critical for transplantation and to categorize immune-mediated events between donor and host posttransplant. Mechanisms will be defined, with details on the contributing cells and factors involved in transplant acceptance versus rejection. Rejection topics will be discussed, including graft-versus-host disease (GVHD). Finally, classes of immunosuppressive agents will be presented to assess therapeutic intervention as a way to control immune features that affect graft acceptance and rejection.

TRANSPLANTATION DEFINED

The concept of organ replacement has become an important part of modern medical therapy. It has been established experimentally that skin can be transferred to different sites on the same person with great success. This is referred to as an **autograft**; all molecules are identical within the individual and the **syngeneic** tissue is recognized as the self. However, tissues transferred between nonrelated individuals (**allograft**) are not readily tolerated; their cellular components are recognized as foreign antigens. Immune responses are initiated within the recipient to eliminate the foreign tissue. Likewise, tissues from nonrelated species

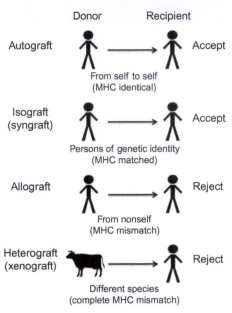

FIG. 11.1   Transplantation acceptance as a function of recipient and donor genetic similarity. Tissue transplantation is governed by immunological rules that allow graft acceptance according to the degree of genetic relatedness between recipient and donor.

(**xenograft** or **heterograft**) share a similar fate and are rejected rapidly unless a high degree of immunosuppression is present. The rules that govern graft acceptance and rejection, as well as the immunological basis of successful graft acceptance, are well defined (Fig. 11.1).

TISSUE HISTOCOMPATIBILITY

The basic architecture of tissues between individuals is quite similar. Indeed, a kidney is a kidney; the liver functions in a similar manner from one person to another. Unfortunately, significant differences in molecules present on the surface of cells exist between genetically different individuals. These discrepancies must be taken into account during the transfer of blood, cells, and tissue. Specifically, serum must be matched to limit interactions with naturally occurring reactive antibodies, blood cells must be matched for carbohydrate markers on their surface, and solid tissues must be matched for overall genetic histocompatibility.

TABLE 11.1 Natural Isohemagglutinins

Blood type	Antigen or antigens present on RBCs	Isohemagglutinin reactivity
A	A	Anti-B
B	B	Anti-A
AB	A and B	None
O	None	Anti-A and Anti-B

NATURAL ISOHEMAGGLUTININS

A subpopulation of immunoglobulin M (IgM) isotype antibodies includes the **natural isohemagglutinins**, which are reactive with the red blood cell (RBC) molecules of the ABO series. The **ABO blood group** epitopes are carbohydrates in nature; antibodies elicited by environmental (bacterial) carbohydrate motifs cross-react with human A or B blood group antigens on RBCs. It is critical, therefore, to match the ABO blood types when giving whole blood or serum. Table 11.1 gives the reactivity of isohemagglutinin antibodies normally found in patients with the various blood groups.

In addition to the ABO antigenic category, an antigen called the *Rh factor (rhesus factor)* is present on blood cells. The Rh factors must be matched; Rh-negative blood is only given to Rh-negative patients. A Rh-negative individual will make antibodies to the Rh factor if Rh-positive blood is given. If antibodies to the Rh factor are present, they will cause agglutination of the donated blood cells. Rh-positive blood or Rh-negative blood may be given to Rh-positive patients, as those individuals do not make antibodies to molecules that they already possess on their own cells.

HUMAN LEUKOCYTE ANTIGENS

The major histocompatibility complex (MHC) antigens are the strongest indicator for inducing allograft rejection. These are the **human leukocyte antigens** (HLAs) discussed in Chapter 4, which allow T cells to recognize presented antigens as a first step in activation events. Relative to immune function, class I HLA molecules (HLA-A, HLA-B, HLA-C) are found on all nucleated cells and mediate the recognition of endogenous antigens by CD8+ cytotoxic T cells. Class II HLA molecules (HLA-DR, HLA-DP, HLA-DR) are on the surface of professional antigen-presenting cells

(APCs), and show exogenous antigens to CD4+ T helper subsets. Subsets of these molecules are inherited from both parents, allowing unique patterns to be expressed in their offspring.

The nature of these molecules includes a high degree of polymorphism, which essentially creates great differences among individuals. As a group, these sets of HLA surface molecules are referred to as **alloantigens**. During transplantation, the histocompatibility alloantigens expressed on donor tissue are recognized by both CD4+ T helper and CD8+ cytotoxic lymphocytes present in the recipient host. The greater disparity between the host and the donor, the greater the lymphocytic reactivity and chance for subsequent tissue rejection are. **Minor histocompatibility antigens**, as well as tissue-specific differences between the host and the donor, may contribute to graft rejection as well.

Tissues transplanted to **immunoprivileged** sites do not typically require MHC matching. For example, corneal transplants do not routinely require HLA matching. The fetus is another example of tolerated non-matched tissue. Although there are common antigens between mother and child, there are also numerous paternal-derived moieties. Factors that allow tolerance include downregulation of MHC on the developing fetus and change in environmental cytokines or factors produced by both the mother and the fetus.

ALLOGRAFT REJECTION

Allograft rejection involves a series of humoral and cellular responses (Fig. 11.2). The immune response involved in allograft rejection spans a wide variety of defined mechanisms. Preformed antibodies can bind to donor tissue, establishing a nidus for direct killing via complement deposition. Antibodies also can function in concert with natural killer (NK) cells in an antibody-dependent cell cytotoxic manner to lyse non-matched target tissue. CD4+ cells recognize class II MHC molecules on the donor tissue (HLA-DP, HLA-DQ, HLA-DR), and are induced to secrete interleukin-2 (IL-2), gamma interferon (IFN-γ), and tumor necrosis factor alpha (TNF-α). These in turn activate CD8+ cells, NK cells, and incoming macrophages. The Th1 CD4+ cells also give signals to activate the Th2 CD4+ group to secrete the cytokines interleukin-4 (IL-4), interleukin-5 (IL-5), and interleukin-10 (IL-10), which can induce B cells to undergo activation and immunoglobulin production, as well as isotype class-switching. The mechanisms involved permit the establishment of rejection categories, including hyperacute, accelerated, acute, and chronic.

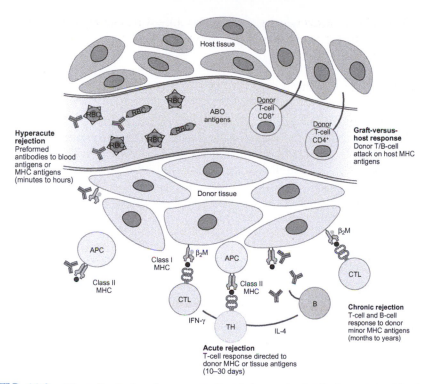

FIG. 11.2 Allograft rejection. Immune-mediated tissue rejection is characterized by the response speed toward donor tissue, which is directly related to immune mechanisms involved in the rejection process.

HYPERACUTE REJECTION

Hyperacute rejection occurs within minutes of transplantation in individuals who are MHC mismatched or in individuals preexposed to the donor's MHC types by prior grafting or blood transfusion. The result is graft tissue loss in a nonreversible manner. The basis relates to preexisting antibodies that react to the mismatch. Natural IgM antibodies are present to cross-reactive epitopes on pathogens that mimic the carbohydrates within components on nonmatched blood cells. These antibodies immediately recognize the foreign tissue and activate complement, which in turn releases factors attracting and activating neutrophils.

ACCELERATED REJECTION

Accelerated rejection, also called "second-set rejection," is relatively rare but occurs with multiple transplants from genetically related donors.

Recipients who have rejected a previous allograft tend to reject a second allograft from the same donor significantly faster.

ACUTE REJECTION

Acute rejection, also called "first-set rejection," occurs between 10 and 30 days after grafting in untreated recipients. This is the expected time for reactive T-cell populations to expand and react. Both T helper and T cytotoxic cells are usually required, although the direct cytotoxic event is delivered by CD8+ cytotoxic T lymphocytes (CTLs). Reactions usually occur later in immunosuppressed recipients, depending on the amount of immunosuppression success.

CHRONIC REJECTION

Chronic rejection occurs months to years after the transplant. It is a complex reaction involving the maturation of both T- and B-lymphocyte responses. Antibodies are directed at the foreign (nonself) antigens within the graft. Subsequent deposition of antibody-antigen complexes leads to targeted destruction of graft tissue and indirect damage to vascular beds. Chronic rejection leads to permanent damage that is difficult (if not impossible) to reverse with immumosuppressants.

The molecular mediators involved in graft rejection are depicted in Fig. 11.3. It is relatively straightforward to envision the acute and chronic mechanisms discussed as being a direct recognition by the host T cell to a combination of foreign MHC and foreign antigens. However, keep in mind that it is also possible for alloantigens to be presented by host APCs. When host cells pick up pieces of the donor tissue and present donor-derived peptides, they also can be targeted for destruction. Indirect recognition of host lymphocytes, therefore, may contribute to the destruction of self-tissue that is physically near the grafted organ.

GRAFT-VERSUS-HOST DISEASE (GVHD)

Graft-versus-host disease (GVHD) occurs when immunocompetent lymphocytes from the donor tissue are inadvertently delivered to the host during the transplantation process. The host, who is immunosuppressed at the time of transplantation, does not reject the alloreactive cells. Over time, these infiltrating donor cells expand, culminating in a pool of donor cells reactive to host tissue. GVHD can occur when there is a mismatch of HLA (class I or class II), or if there are a significant number of differences in minor histocompatibility antigens, such as that seen in closely matched siblings. A common GVHD occurrence is after **bone marrow transplantation,**

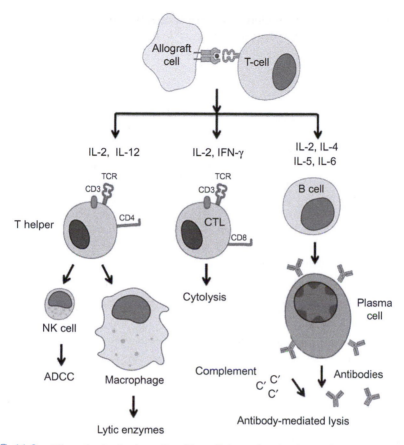

FIG. 11.3 Allograft rejection is mediated by cellular and molecular mediators. The specific molecular mediators of allograft graft rejection are a function of the responding immune cell phenotypes involved in tissue recognition.

a procedure where alloreactive hematopoetic stem cells are delivered as a form of gene therapy. These pluripotent stem cells give rise to all phenotypes of white blood cells (WBCs); however, the pool of transplanted cells often contains mature lymphocytes that are capable of recognizing differences in HLA between the donor and host. Stringency of T-cell depletion prior to transplantation reduces this occurrence.

PRETRANSPLANTATION HISTOCOMPATIBILITY EVALUATION

Multiple laboratory methods are used to evaluate tissue histocompatibility between donor and recipient. Methods allow the matching of tissues between individuals, with a higher degree of graft survival directly related to level of similarity. Any donor-recipient HLA incompatibility can result in an immune response, rejection, and possible graft loss. And while immuno-suppressants may obviate the impact of HLA matching for both short- and long-term graft outcome, it is preferred to limit mismatch prior to transplantation.

Any potential donor must undergo extensive screening prior to trans-plantation. This begins with a test for mismatch blood cell Rh-antigens between host and donor, as well as direct testing for host antibody reac-tivity to donor target cells. In essence, this simple test can identify reactive antidonor antibodies by examining the level of cytotoxicity and lympho-cytotoxicity of these antibodies to lyse cells in the presence of complement components. The purpose of the cross-match is to detect clinically relevant IgG antidonor antibodies to prevent hyperacute, accelerated, or chronic rejection. The next level of tests examines cellular reactivity, accomplished using host and donor T cells in a **mixed lymphocyte reaction** to assess direct reactivity to allogeneic MHC between individuals. Basically, if T cells are reactive to allogeneic molecules, they will undergo rapid rep-lication and produce diagnostic secretion of cytokine subsets.

Recent technological advances now permit identification of haplotype distinctions between individuals without cell culture methods. These methods are especially useful when comparing relationships of parents and siblings to the recipient. One particular method, called HLA-DNA (also called *PCR typing*), uses polymerase chain reactions (PCRs) and sequence-specific oligonucleotide probes to identify DNA-genomic sub-types rapidly. DNA primers can be used that are specific for individual or similar groups of HLA alleles, allowing the amplification of relevant genomic DNA. This type of screening is especially useful when donor tissue is of cadaveric origin (i.e., from a deceased person).

IMMUNOSUPPRESSIVE DRUGS TO PREVENT ALLOGRAFT REJECTION

At the present time, there is no successful clinical protocol to induce complete tolerance of allografts. Realize that the act of replacing an organ is by definition a trauma, where inflammation is induced through surgical intervention; inflammation is key to priming innate reactivity to present donor antigens to the adaptive arm of the immune response. All patients require daily, lifelong treatment with immunosuppressive agents to inhibit graft rejection. All immunosuppressive agents used in clinical practice have drawbacks relating to toxicity and side effects or to failure to provide sufficient levels of downregulated lymphocytic response. On one hand, inadequate immunosuppression allows the recipient to mount an immune response, causing allograft rejection. On the other hand, excessive immunosuppression can lead to the development of opportunistic infections and neoplasia.

IMMUNOSUPPRESSIVE THERAPY

Immunosuppressive agents are often used to control reactions prior to graft rejection. These agents fall into categories that depend on the targeted function and immune modification desired (Table 11.2). Prior to transplantation, agents are given to the recipient at a relatively high level to quiet immune reactivity, allowing greater success and acceptance of tissue immediately posttransplantation. After the tissue has been placed

TABLE 11.2   Immunosuppressant Drugs and Therapeutics

Class	Mechanism of action	Example
Corticosteroids	Blocks the expression of multiple cytokines	• Prednisone • Prednisolone • Methylprednisolone
Cytotoxic agents	Blocks DNA synthesis or replication in proliferating cells; can suppress APC processing	• Azathioprine • Cyclophosphamide • Hydroxychloroquine • Methotrexate
Immunophilin ligand	Blocks T-cell activation and gene transcription	• Cyclosporine • Tacrolimus (FK506)
Proliferation signal inhibitors	Block intracellular kinases	• Sirolimus (rapamycin) • Mycophenolate mofetil
Immunosuppressive monoclonal or polyclonal antibodies	Specifically target adaptive cells and cellular functions	• Antilymphocyte globulin • Antithymocyte globulin • Anti-CD3 MAb (OKT3) • Rituximab (anti-B cell) • Rh(D) immune globulin • Etanercept (anti-TNFα/β) • Daclizumab (anti-IL2R) • Abatacept (anti-CD28/CD80)

within the host, the major concern revolves around targeting the immune system in a manner that prevents reactivity. Maintenance therapy is usually given at a low level to keep the immune system operational but quiet, and without completely shutting down reactivity to opportunistic infections.

It is only when clinical symptoms arise indicating the initiation of active rejection that specific and aggressive immune suppressants are used. In this case, targeted therapeutics are administered to support mechanisms that disrupt immune events and even kill rapidly expanding lymphocytes that demonstrate reactivity to the donor organ. Examples of targeted therapeutics include those aimed at control of critical metabolites, inhibition of T-cell function, and blocking the activity of critical cytokines, either directly or through limiting interaction with their cell surface receptors.

Immunosuppressive therapy has had a significant impact on both the prevention and treatment of rejection. Yet suppressing the immune response has consequences, such as increased risk of infection and certain types of malignancies. While steroids remain an important immunosuppressive clinical tool, recent advances in protocol development limit their use to minimize known side effects. Other therapeutics, such as Tacrolimolus, are effective; however, a major concern about nephrotoxicity remains. Targeted agents, including polyclonal and monoclonal antibodies (biologics), are becoming increasingly useful in the arsenal against rejection of transplanted tissue. However, they often leave the recipient highly susceptible to infection, which remains a major cause of mortality posttransplantation.

SUMMARY

- The immunological rules for transplant acceptance or rejection are governed by recipient responses to histocompatibility molecules on donor cells.
- Allograft reactivity, as well as the speed of rejection, are governed by cell phenotypes and molecules involved in reactivity to donor histocompatibility antigens.
- GVHD represents a state in which immune-competent cells from the donor tissue escape initial destruction, leading to subsequent reactivity against recipient tissues.
- Modern laboratory techniques can use genetic sequences to identify potential histocompatibility mismatches. Therapeutics have evolved to target specific immune responses detrimental to graft acceptance.

12

Assessment of Immune Parameters and Immunodiagnostics

CHAPTER FOCUS

To provide an overview of in vitro antibody binding to antigens that allows the development of successful immunoassays. The information presented will compare and contrast past and current methods of immune detection. There also will be a discussion of commonly used assays to determine cell function in the clinical setting and an introduction to concepts behind large-scale data collection and analyses used in the laboratory.

ANTIBODY-ANTIGEN REACTIONS

The unique structural characteristics of antibodies (or immunoglobulins) can be exploited for use in the laboratory. Advances in molecular design related to monoclonal antibody engineering, protein biochemistry, and recombinant DNA technology have allowed continued refinement of applications for use in clinical practice and in cutting-edge experimental research. The requirements for antibody-antigen reactions are discussed next.

AFFINITY

The reaction of an antigen with its homologous antibody, and the subsequent physical manifestation of that binding, constitute a two-stage phenomenon. The initial (or primary) binding reaction between the complementarity-determining region (CDR) domains on the Fab portion of the **antibody (Ab)** and the **antigen (Ag)** occurs fairly rapidly and invisibly.

The interaction of the CDR of the antibody with its cognate antigen is not covalent. Instead, binding is mediated by van der Waals forces, electrostatic interactions, and hydrophobic interactions (Fig. 12.1). Hence, Ab-Ag binding interactions are analogous to those observed in enzyme-substrate reactions and can be defined similarly through physical laws of mass action. The rate at which the antibody forms the Ab-Ag complex is referred to as its **affinity**, or strength of binding, while the dissociation constant represents the stability of the Ag-Ab complex. Both effective binding and stability contribute to the strength of an Ag-Ab interaction. Hence, an antibody with a low affinity but a higher dissociation constant is more effective at binding antigen than an antibody with a high affinity constant but a smaller dissociation constant. Strength of binding is also influenced by antibody isotype. Immunoglobulin G (IgG) antibodies

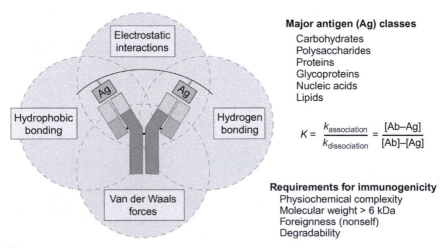

Major antigen (Ag) classes
Carbohydrates
Polysaccharides
Proteins
Glycoproteins
Nucleic acids
Lipids

$$K = \frac{k_{association}}{k_{dissociation}} = \frac{[Ab\text{–}Ag]}{[Ab]\text{–}[Ag]}$$

Requirements for immunogenicity
Physiochemical complexity
Molecular weight > 6 kDa
Foreignness (nonself)
Degradability

FIG. 12.1   Forces contributing to antibody-antigen interactions. The interactions between antibodies and antigens follow the laws of mass action and can be described in a representative equation. K represents the equilibrium constant; $k_{association}$ is the association constant, $k_{dissociation}$ is the dissociation constant, [Ab] represents the free antibody concentration, [Ag] represents the free antigen concentration, and [Ab–Ag] represents a complexed antibody with an antigen.

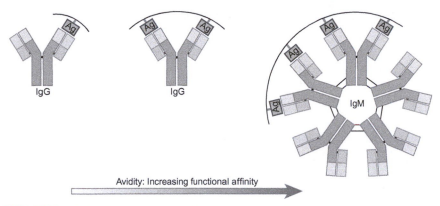

FIG. 12.2   Multivalent interactions of antibody-antigen binding. Avidity represents functional antibody-binding to the antigen, increasing as more epitopes are bound.

typically have a higher affinity for antigen than immunoglobulin M (IgM) antibodies. However, the pentameric form of IgM allows it to bind multiple antigens (5–10 paratopes in IgM versus 2 paratopes in IgG). Thus, IgM has a higher **functional affinity** (**avidity**) than IgG (Fig. 12.2).

SECONDARY MANIFESTATIONS OF ANTIBODY-ANTIGEN BINDING

Cross-linking of Ag and Ab depends on several factors, including the isotype and specificity of the antibody and the size and epitopes (unique conformations) contained within the antigen. This reaction is affected by the number of binding sites available for each Ab and the maximum number of binding sites on an Ag or particle. This concept is defined as the **valence** of the antigen or antibody; the valence of Ab and Ag has to be ≥2 for precipitation to occur. It is important to note that steric considerations limit the number of distinct Ab molecules that can bind to a single antigen at any one time.

The isotype of antibodies, determined by the expression of specific heavy chains (discussed in Chapter 3) plays an important role in cross-linking complex antigens. The flexible hinge region of IgG, for example, allows this antibody to bind two identical antigens simultaneously. The hinge region of IgM is not as flexible as IgG, and yet its pentameric form allows it to bind multiple antigens at once. The number of antibody-binding sites, or epitopes, on an antigen affects cross-linking as well, as do the physical properties of the antibody contact points (called the **paratope**) that interact with the epitope. As shown in Fig. 12.3, the

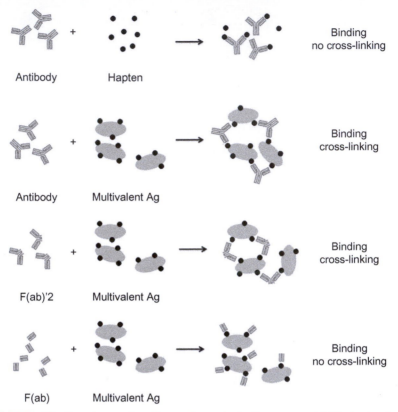

FIG. 12.3 The Coombs reaction dictates the agglutination process. Multivalent antigens are easily precipitated with polyclonal antibodies, but not when monoclonals are used. As each antibody-binding complexity is reduced, the ability to form precipitates depends on the complexity of the antigenic structure.

presence of multiple epitopes recognized by the same antibody (**unideterminant**) or multiple epitopes recognized by different antibodies (**multideterminant**) on an antigen increases the possibility of cross-linking and precipitating the particle. In contrast, antibody-binding to very small antigens, called **haptens**, cannot cross-link. Finally, Ag-Ab cross-linking is more likely to occur if the antibody in solution is **polyclonal** (i.e., it is comprised of multiple immunoglobulins with many different antigen specificities). It is noteworthy that normal human serum samples contain polyclonal antibodies with many different specificities; therefore, agglutination and precipitation reactions occur in vivo naturally.

In both agglutination and precipitation reactions, various amounts of soluble antigen are added to a fixed amount of serum that contains antibody. As illustrated in Fig. 12.4, when small amounts of Ag are added,

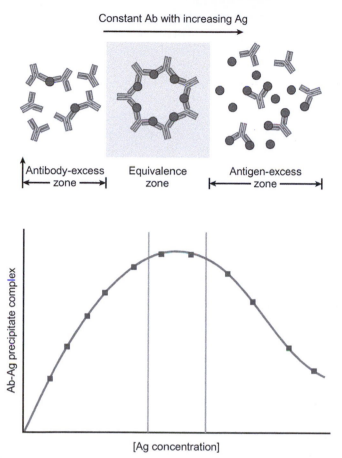

FIG. 12.4 Physical interactions dictate antibody-antigen interactions. The amount of precipitate formed between antibody and antigen is a result of relative concentrations of excess versus equivalence.

Ab-Ag complexes are formed with excess Ab, and each molecule of Ag is bound by Ab and cross-linked to other Ab molecules. No cross-linking can occur due to this **prozone effect**. When enough Ag is added, all the Ab and Ag complexes fall out as precipitate (**zone of equivalence**). When an excess of Ag is added, only small Ag-Ab complexes form (no cross-linking) and the precipitate is reduced. The highest dilution of serum that induces agglutination, but beyond which no agglutination occurs, is called the **titer**; this measurement is often used to compare the relative amount of Ab binding Ag in clinical samples.

   Agglutination reactions occur when polyclonal antibody binds to multideterminant particulate (insoluble) antigens. A very good example of a commonly used agglutination assay is red blood cell (RBC) typing. RBCs express ABO antigens, which can be recognized by IgM antibodies that are very efficient at cross-linking. Therefore, serum from a person who expresses the O phenotype will contain antibodies specific for A and B blood group antigens. Incubation of either type A or type B blood with this serum will result in binding of the endogenous IgM antibodies to the RBCs, and agglutination, or clumping due to Ab-Ag cross-linking, will occur. Conversely, type O blood cells will not be recognized by serum from type A or type B donors. Hence, type O blood is a so-called universal donor phenotype, while patients with AB phenotype are so-called universal acceptors.

Due to the presence of sialic acid on the membrane, RBC can possess a net negative charge. In solution, this charge, called the **zeta potential**, can prevent cross-linking. This is best illustrated by the inability of IgG to cross-link RBCs, while the larger structure and multiple binding sites of the IgM antibody can overcome this potential. In some cases, however, it is necessary to detect IgG binding to RBC. The **Coombs reaction** is a widely used agglutination reaction that is useful for the detection of maternal IgG antibodies directed against Rh+ antigens found on the fetal RBC.

In the **precipitation reaction**, Ab binds to soluble Ag. Cross-linking of multivalent Ag by divalent Ab forms a lattice structure linking the smaller Ag complexes. When the lattice grows to a large-enough size, it loses solubility and precipitates out of the solution. Precipitation reactions are governed by the same rules as agglutination reactions; lattice formation will occur only in the zone of equivalence and will be inhibited either by Ab excess or Ag excess. Soluble precipitation reactions are performed by adding various concentrations of Ab to a constant concentration of Ag (or the reverse) and performing serial dilutions to determine the titer. Hence, these assays may be considered semiquantitative, as they can determine the relative quantity of Ab in one sample versus another.

Nephelometry is a widely used method for accurately measuring precipitated quantities of immunoglobulin classes in serum. In this assay, proteins in the sample react with specific antibodies to form particulates. As light passes through the aggregated suspension, a portion of the light is scattered and measured for comparison against stored standards. Thus, this is a quantitative method using liquid-phase precipitation principles.

SOLID-PHASE PRECIPITATION ASSAYS

To generate a simple yes/no answer to the question of whether a specific Ag or Ab is present in a sample, it can be helpful to slow the rate of diffusion in a gel matrix and immobilize precipitates for subsequent visualization, either directly or with the aid of various staining methods. Several qualitative and quantitative methods are in wide use in medicine today for the analysis of numerous hormones, enzymes, toxins, and the products of the immune system itself.

RADIAL IMMUNODIFFUSION

In this reaction, a known antibody or an antigen is infused into the gel matrix. A test sample is placed in the center of the gel. As the unknown sample diffuses into the surrounding agar, a precipitation reaction will occur if there is a positive Ab-Ag interaction (Fig. 12.5). Because precipitation happens only at the zone of equivalence, a ring will form some distance away from the high concentration of antigen at the center. The radial immunodiffusion can be a quantitative assay if the diameter of rings formed from various known quantities of antigen is used to generate a standard comparison curve.

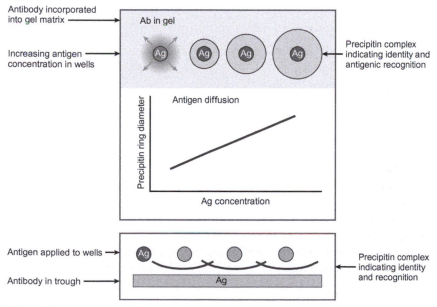

FIG. 12.5 Diffusion assays. The radial diffusion assay *(top)* is characterized by precipitin reactions occurring as an applied antigen diffuses through a gel matrix containing reactive antibodies. The diameter of the precipitin ring formed can be quantitated by comparison to known standard antigen concentrations. A variant on the radial diffusion assay *(bottom)* allows qualitative determination as antibodies and antigens diffuse toward each other through the assay matrix.

THE OUCHTERLONY ASSAY

The Ouchterlony Assay, developed by Orjan Ouchterlony in the 1950s, allows comparison to determine the relatedness of two antigens. The assay is called a **double diffusion** assay because both the antigen and antibodies are diffusing. It is a qualitative assay—a reaction occurs, or it doesn't. Antibodies and antigens are placed in separate but close wells. The molecules then diffuse slowly into the agar in a radial fashion toward each other. A positive, thin, opaque line will form in the agar at right angles to a line connecting the centers of the two wells if precipitation occurs.

IMMUNOELECTROPHORESIS

Immunoelectrophoresis is a variation of the Ouchterlony double diffusion assay in gel technique, designed to analyze complex protein mixtures containing various antigens. Electrophoresis first separates proteins according to their size and mobility in the electric field within a gel matrix. Next, a mix of antibodies specific for the proteins is added to a trough cut in the agar. The individual proteins and their specific antibodies will diffuse toward one another, and lines of precipitate form, representing interactions. The medical diagnostic use is of value when certain proteins are suspected of being absent (e.g., hypogammaglobulinemia) or overproduced (e.g., multiple myeloma).

LATEX AGGLUTINATION

The latex agglutination reaction takes advantage of clumping observed when a sample containing the specific antigen is mixed with an antibody that has been coated onto latex particles. The use of synthetic latex beads offers the advantages of consistency and uniformity to the laboratory. The latex agglutination assay offers a rapid advantage, with results sometimes generated within minutes. Composition of synthetic beads can vary greatly, allowing binding of chemically complex antigens in small clinical samples.

LATERAL FLOW

The lateral flow assay is a paper-based platform used in the clinical laboratory for the detection and quantification of molecules from within a complex fluid mixture. A microstructured support polymer is used in the transport of fluids. It serves as both a sponge to hold the sample, and then as a matrix for solute migration to a secondary conjugate pad with bioactive particles. An optimized chemical reactivity occurs between a target antigen and an immobilized antibody. Antigens bind to particles while they travel with the fluid through reactive zones. A third level of captures the antigen:antibody complex. The Lateral Flow assay also can function in a manner to capture antigens by sandwiching them between antibodies directed toward multiple unique antigenic determinants on the molecule.

WESTERN BLOT

The mechanisms that underlie immunoelectrophoresis form the basis for the more commonly used western blot (also called an *immunoblot*). A mixture of antigens is first separated by size via electrophoresis on a gel, after which the proteins are transferred onto a solid matrix such as nitrocellulose that tightly binds proteins. Next, patient serum containing antibodies suspected of binding to the antigen can be added, followed by antihuman detection antibodies that have an enzyme covalently attached. Substrate for the enzyme is added and turns colors when enzyme is present, and the colored line determines the detection of antigen if it is present (Fig. 12.6). Importantly, since western blotting separates proteins by size before they are detected by antibodies, it is possible to identify several

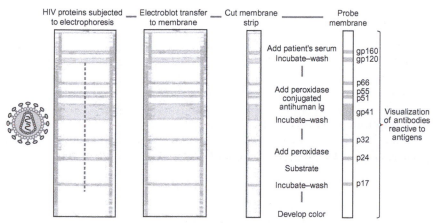

FIG. 12.6   Western blot (immunoblotting). Reactivity to antigens assessed by western blot analysis using serum antibodies is pictured. HIV antigens separated by gel electrophoresis are transferred to a solid matrix (nitrocellulose), and then probed with patient serum. Reactive antibodies against HIV antigens are visualized, thus determining patient exposure to the viral agent.

specific antigens in one sample. A good example of this is the test used to detect reactivity to human immunodeficiency virus (HIV)—serum from patients who are infected with this virus usually contains antibodies that bind to multiple proteins (GAG, POL, and ENV), resulting in the visualization of multiple reactive lines.

IMMUNOASSAYS

The methods discussed thus far in this chapter rely on the physical manifestation of binding of polyclonal Abs to multivalent, multideterminant Ags. The heart of the immunoassays lies in the inherent ability of antibodies to recognize specific, unique target antigens. In the laboratory, this is essential for the clinical detection of pathogenic molecules. As a general rule, immunoassays utilize methods with great sensitivity and specificity to detect antibody-binding to antigens, with direct or indirect labeling of the antibody-constant region. Many immunoassays rely heavily on the use of monoclonal antibody preparations to bind to known antigenic epitopes.

Solid-phase immunoassays are a group of assays in which the antigen or the antibody is coated on the surface of a plastic microplate and sensitive indicators (radioactivity, enzymatic action, or fluorescence) are used to detect the presence of Ag or of Ab. These assays may be further characterized according to the types of antigens being analyzed: soluble or cellular.

ENZYME-LINKED IMMUNOADSORBENT ASSAY (ELISA)

The **enzyme-linked immunoadsorbent assay (ELISA)** is solid-phase assay in which a detection antibody molecule is coupled to an enzyme that converts added substrates to a colored product for spectrophotometric detection (Fig. 12.7). A very common enzymatic label is horseradish peroxidase (HRP), which, when incubated with a peroxidase substrate such as tetramethylbenzadine (TMB), results in a blue solution that is detected at 650 nm.

In general, two types of ELISA assays are used in the clinic and in biomedical research. The first, a direct-binding ELISA, uses an antigen bound to a plastic plate. A sample antibody is incubated and allowed to bind, and detection occurs via a secondary-labeled antibody. A second type of ELISA, the sandwich or capture ELISA, uses a pair of specific sets of unique antibodies to bind antigen. The primary antibody is coated on the plate, and then a sample containing antigen is added.

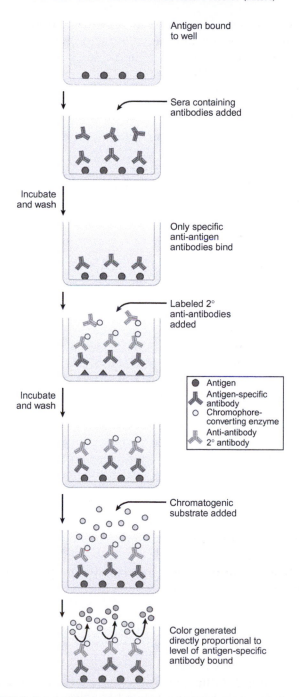

FIG. 12.7 ELISA. Serum added to wells coated with antigen is probed with antibodies. Enzyme-conjugated secondary antibodies detect the primary complex. Catalytic conversion of chromogenic substrate allows quantitation against known standards.

Antigen binds to the coated antibody, and then a labeled secondary antibody that binds to a different epitope on the same antigen is added to create a tiered complex. Color change indicates binding, and thus the presence of specific antigen in the sample.

A variation on the ELISA method is the **ELISPOT assay**. In this assay, a nitrocellulose membrane is substituted for the solid plastic surface of the well. Primary antibodies or antigens are bound to this membrane, and then incubated with living cells in a solution. The living cells secrete a product (e.g., an antibody or cytokine), which is subsequently captured on the membrane. Labeled secondary antibody is added, and reactivity is determined through the use of a precipitable substrate that forms detectable spots on the membrane. Then the number of cells secreting the protein of interest may be enumerated.

DETECTION OF CELLULAR ANTIGENS

It can be of great diagnostic value to determine if a particular antigen is found on or within the cells of a particular tissue. Assays can be performed directly on biopsies of tissue and visualized using a microscope. The immunofluorescence method utilizes the covalent attachment of fluorescent organic compounds to specific antibodies, which then can be used to detect antigen in the tissue sample. The fluorescent compounds excite at different wavelengths and thus can be detected using special microscope lighting.

This is a highly sensitive and specific assay, and individual cells can be stained with up to 12 excitable compounds. Visualization is accomplished by **direct immunofluorescence**, in which an antibody specific for the antigen in question is directly labeled with the fluorophore and used to identify the antigen. Or it may be accomplished by **indirect immunofluorescence**, in which a two-step method in which the unlabeled antibody specific for the antigen in question is reacted first, and then the slide is flooded with a fluorescent molecule to detect the bound antibody. Immunofluorescence is used in clinical laboratories to screen for anti-DNA antibodies in suspected cases of autoimmune systemic lupus erythematosus.

IMMUNOHISTOCHEMISTRY

Immunohistochemistry is a powerful technique to localize antigens on tissues embedded in paraffin and placed on glass slides. Incubation with specific enzymatically labeled antibodies allows the detection of molecules within the tissues; detection is enhanced using enzymes, such

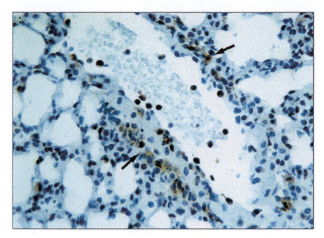

FIG. 12.8 The immunohistochemical staining method. Antibodies can identify cell structure and phenotype within tissue. Pictured here are lymphocytes cuffing pulmonary vascular regions *(arrows)*. Enzyme (HRP)-conjugated antibodies directed against surface CD3+ were incubated with the tissue section. Conversion and deposition of chromogenic substrate (diaminobenzidine) enables visualization of T lymphocytes *(brown stain)*.

as HRP or alkaline phosphatase, for which addition of substrate then colorizes the membranes of the cells expressing the antigen of interest (Fig. 12.8). While this method is usually less sensitive than immunofluorescence, it is often used in clinical pathology laboratories to permit determination of antigenic placement within overall general tissue morphology.

FLUORESCENCE-ACTIVATED CELL SORTING (FACS) ANALYSIS

Fluorescence-activated cell sorting (FACS) analysis is used to identify, and sometimes purify, one cell subset from a mixture of cells (Fig. 12.9). This is an extremely effective tool to identify and/or isolate specific cell subsets, as it allows rapid identification, as well as quantification of cells expressing specific surface molecules. Antibodies directed against surface molecules (such as CD surface proteins) may be directly tagged with fluorescent compounds or indirectly identified with secondary tagging (indirect immunofluorescence). Labeled cells are passed single-file through a laser beam by continuous flow in a fine suspension stream. Each cell scatters some of the laser light. Detection of scattered light in a parallel manner (forward scatter) is indicative of cell size, whereas detection of scattered light at a perpendicular angle (side scatter) is indicative of cellular granularity.

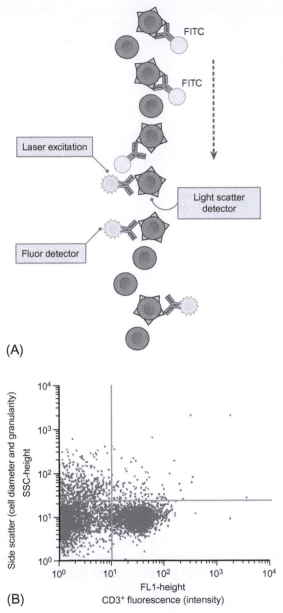

FIG. 12.9 Fluorescence-activated cell sorting. Flow cytometric analysis allows direct identification of cell phenotypes using fluorescently tagged antibody molecules. Modern cytometers can measure multiple parameters simultaneously, including low-angle forward scatter (proportional to cell diameter), orthogonal scatter intensity (measuring cellular granularity), and fluorescence intensity at several wavelengths (A). Pictured here is a scatter-gram of splenocytes screened for reactivity with a fluorescein isothiocyanate (FITC)-labeled anti-CD3 antibody to quantitate T lymphocytes (B). A shift to the right on the *x*-axis indicates the presence of positively stained cells with increasing levels of CD3 surface molecule. *SSC*, side scatter.

These physical parameters alone can subdivide granular polymorpho-nuclear cells (PMNs), mononuclear cells, and RBCs. Flow cytometers are equipped with multiple lasers to generate different excitation wavelengths, as well as a series of optical detectors that capture the unique emission wavelengths generated by the various fluorescent labels. Thus, specific monoclonal antibodies labeled with unique fluorescent compounds may be used to detect up to 20 unique antigens on a single cell.

Newer methodologies also allow the detection of intracellular cytokines to match surface molecules found on individual cell phenotypes. The instrument can identify the cells of interest based on size or fluorescent label. Downstream of the laser beam, an electrostatic deflection system within the instrument, can sort the cells of interest into batches. Sorting of cells also can be accomplished using antibodies coupled to magnetic beads. Then the cells are placed over a magnetized column and labeled cells can be isolated from the nonlabeled population.

MULTIPLEX BEAD ARRAYS

A relatively new technology combines aspects of an ELISA with the sensitivity of flow cytometry. **Multiplex bead arrays** rely on the engineering of microspheres internally coded with two fluorescent dyes. Combinations of the dyes can be used to generate up to 100 individual bead sets, each of which can be coated with a specific antibody. Bead sets may be incubated with samples (plasma or cell culture supernatant) in a single tube, allowing the measurement of multiple parameters at once. The antibody-coated beads bind to their specific antigen target, and then biotin-labeled secondary detection antibodies are used to "sandwich" the antigens bound on the beads. A reporter molecule, such as streptavidin-phycoerythrin, then indicates each complex. In a method similar to that used in flow cytometry, the beads are passed single-file through a laser beam; complexes are identified by fluorescence with internal fluorochromes unique to each bead, emitting specific signals detected by digital processors. In this manner, multiple analytes can be detected in a single sample, making this a powerful tool for screening patient responses in various disease conditions.

ASSAYS TO DETERMINE IMMUNE FUNCTION

Assessment of immune function is a useful clinical tool to establish competency, as well as defects in the patient. The tests given next allow the indication of immune disorders with identification of response for excesses or deficiencies.

COMPLEMENT FIXATION TEST

The complement fixation test is a blood test that can determine the presence of antigen-specific antibodies by incubating patient serum with antigen and complement. This assay takes advantage of the requirement for complement to be activated by the combination of antigen-antibody complexes (Fig. 12.10). Should specific complexes form, the complement cascade will be activated. A target for this activation is thus required to determine reactivity. Sheep RBCs (sRBCs) bound to anti-sRBC antibodies are often used as an activation target.

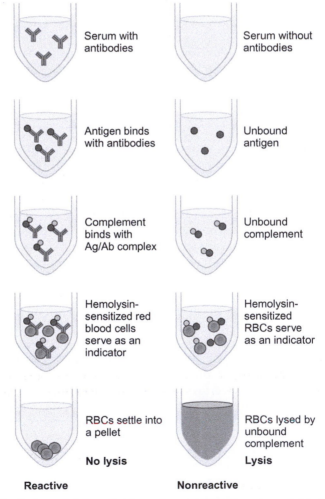

FIG. 12.10   In the complement fixation assay, complement components bind to antibody-antigen complexes, thereby making complement unavailable for the hemolysis of indicator RBCs. In the absence of specific antibody-antigen interactions, complement assembly results in cell lysis.

In a positive test, the complement is bound to an antigen-antibody complex and thus is sequestered away from interaction with target sRBC-Ab complexes. The RBCs remain unlysed and settle to the bottom of a concave-shaped well. In a negative or nonreactive test, the complement remains free to interact with the sRBC-Ab complexes, causing them to lyse. The complement test is a powerful tool to confirm the presence of antibodies following exposure to a specific microorganism. For example, the Wasserman reaction is a diagnostic complement fixation test to detect antibodies to the syphilis-causing organism *Treponema*; a positive reaction indicates the presence of antibodies, and therefore syphilis infection.

LYMPHOCYTE FUNCTION ASSAYS

Lymphocyte function can be compromised in certain diseases or can occur as a result of a genetic abnormality. A diagnosis can be confirmed in many cases if it is known whether the B or T cells are normal, if the existing B cells can make antibodies, or if the T cells can produce the correct cytokines. For example, cytotoxicity assays measure the ability of cytotoxic T cells or NK cells to kill radioactive target cells that express a specific antigen for which the cytotoxic T cells may be sensitive.

Blast transformation assays are used to measure the cell reactivity of unique phenotypic populations within peripheral blood lymphocytes. Cells are incubated in the presence of specific antigens or mitogens. A **mitogen** is any agent capable of stimulating cellular activation. For example, **lipopolysaccharides** can cause polyclonal stimulation of B cells. Several **lectins**, including **concanavalin A** and **phytohemagglutinin**, are effective T-cell mitogens. And **pokeweed mitogen** stimulates polyclonal activation of both B and T cells. If reactive, lymphocytes will proliferate. Historically, radioactive nucleotides (^{3}H-thymidine) were added; the amount of radioactivity incorporated into DNA was determined as a quantitative measure of proliferation. The reduction of tetrazolium salts is now recognized as a safer and accurate alternative to radiometric testing; a yellow tetrazolium bromide derivative salt (referred to as MTT) is reduced in metabolically active cells to form insoluble purple formazan crystals, which then can be quantified by spectrophotometric means.

OTHER TOOLS TO MEASURE IMMUNOLOGICAL STATUS

Monoclonal Antibodies

Polyclonal antibody mixtures consist of antibodies that are specific for a multitude of different epitopes on even simple antigens. Subpopulations of antibodies with different affinities exist even in the subset that is specific for a single epitope. Due to cross-reactivity of antibodies and the need for more controllable assays, it is sometimes of great advantage to have a homogeneous antibody preparation that is specific for a single epitope and with high affinity. In 1975 George Kohler and Cesar Milstein developed a method for making antibodies that are **monoclonal** (i.e., all antibodies are derived from a single precursor plasma cell so that all antibodies are identical; MAb). Genetic-engineering methodologies now allow the easy and rapid production of monoclonal antibodies, with swapping of human sequences for homologous protein utility.

Microarrays

Levels of expression of thousands of genes can be measured simultaneously using **gene chips** or **microarrays**. Briefly, thousands of short complementary DNAs (cDNAs) representing genes from all parts of the genome are attached to a silicon chip. Samples of messenger RNA (mRNA) from cells in culture are reverse-transcribed into cDNA, labeled with different fluorochromes, and used to measure differential expression of distinct genes. The chips are exposed to laser light, and the unique wavelengths emitted by the fluorochromes are measured and compared to control samples. Hence, the relative binding of the cDNAs to each unique sequence can be determined.

Finally, recent advances in sequencing technology and computing algorithms have allowed the development of methods to rapidly screen and characterize polyclonal immune responses to antigen. Collectively, these methods are referred to as **high-throughput Ig sequencing (Ig-Seq)** technologies, and they take advantage of the unique manner in which Ig genes recombine to specifically amplify heavy- and light-chain variable sequences isolated from class-switched B cells or T-cell receptors (TCRs) from memory T cells. Genomic DNA isolated from a population of B cells is amplified using primers complementary to rearranged V(D)J sequences.

Alternatively, cDNA can be amplified using a primer pool that is complementary to leader peptides or framework regions of V-gene segments combined and specific primers for heavy or light chains. Then the samples are sequenced; bioinformatic approaches are used to generate output data. The results from these analyses can be applied to many experimental and clinical questions to understand the generation of responses and their role in human health and disease.

SUMMARY

- The affinity of the antibody-antigen reaction is defined through physical laws of mass action.
- Understanding the nature of antibody-antigen reactions has allowed the development of specific assays to detect functional immune response.
- The immunoassays have become strong clinical tools to determine immune reactivity and allow the assignment of developmental status in individuals with immune disorders.

Glossary

Accessory molecule Cell surface molecules that participate in cellular interactions to modulate the strength and direction of a specific immune response.

Acquired immunodeficiency syndrome (AIDS) An infectious disease caused by the human immunodeficiency virus (HIV) characterized by the loss of CD41 T helper lymphocytes.

Acute phase protein Any of the nonantibody proteins found to be increased in serum during active and immediate innate responses; includes complement factors, C-reactive protein (CRP), and fibrinogen.

Acquired immunity/adaptive immunity A network of antigen-specific specialized lymphocytes that function to eliminate or prevent systemic infection. Responses take days or weeks to develop and result in immune readiness (memory) that may be sustained for long periods.

Adjuvant An excipient added to an immunogen to direct immune responses during vaccination.

Affinity The binding strength of an antibody for its cognate antigen.

Affinity maturation The process by which B lymphocytes mature to increase the specificity of an antibody for its cognate antigen.

Allelic exclusion The expression of only one gene, while the alternative copy (allele) remains silent.

Allergy A misdirected hypersensitive immune reaction to normally harmless foreign substances. An "allergen" is any antigen that elicits Type I hypersensitivity (allergic reactions).

Allogenic Being genetically different from a similar species member.

Allograft A tissue graft from a nonself donor of the same species.

Anaphylactic shock A systemic allergic reaction to circulating antigens, resulting from interaction with immunoglobulin E (IgE) antibodies on connective tissue mast cells, followed by the release of inflammatory mediators that confer the shock.

Anaphylotoxin A complement system enzymatic fragment (e.g., C3a, C4a, C5a) that mediates host defense functions, including chemotaxis and the activation of cells bearing fragment receptors; causes enhanced vascular permeability and mast cell histamine release.

Antibody A surface protein on B lymphocytes that can be secreted in large amounts in response to an antigen. Five subclasses exist, each of which has a unique function to confer protection against infectious assault. *See also* Immunoglobulin.

Antibody-dependent cell cytotoxicity (ADCC) A cytolytic process directed toward an antibody-coated target cell via mechanisms whereby effector cells [mostly natural killer (NK) cells] with Fc receptors recognize the constant region of target-bound immunoglobulin.

Antigen A foreign substance capable of eliciting an immune response; may be protein, carbohydrate, lipid, or nucleic acid in nature.

Antigen presentation An organized display of processed antigenic fragments bound to presenting cell surface histocompatibility molecules to allow targeted recognition by the T-cell receptor (TCR).

Antigen-presenting cell (APC) A specialized bone marrow-derived cell, bearing cell-surface class II major histocompatibility complex molecules to function in antigen processing and presentation to T cells.

Antigen receptor A specific antigen-binding molecule on T or B lymphocytes; comprised of amino acids produced from genetic sequences with physical rearrangements of V-, D-, and J-gene subsets.

Antigenic A substance capable of recognition by an immunoglobulin or an antigen receptor. The *antigenic determinant* is the site or epitope on a complex molecule recognized by an antigen receptor [an antibody or a T-cell receptor (TCR)]. The antigen-binding site *paratope* represents the physical location on the receptor that contacts the molecules.

Apoptosis The process of programmed cell death (PCD).

Autograft Tissue grafted from one person to another place on his or her body, with a complete match of histocompatibility molecules.

Autoimmune disorder The pathological condition in which the body's own immune system is directed toward self-antigens. *Autoreactivity* describes immune cells mounting a response against self.

Avidity The combined strength of antibody-antigen interaction, taking into account multiple binding sites between molecules.

BALT See MALT.

Basophil A polymorphonuclear granulocytic cell involved in allergic reactions during Type I hypersensitivity.

B cell/B lymphocyte The type of lymphoid cell produced in the bone marrow from lymphoid progenitor stem cells that possesses specific antibody cell-surface antigen receptors; a cell capable of producing antibodies when activated.

Blood group antigen A red blood cell (RBC) surface molecule that is detectable with antibodies produced by sensitization to environmental substances. Major blood group antigens include ABO and Rh (Rhesus) markers used in routine blood screening to designate blood types.

CD3 complex A set of signal transduction molecules that assist in T-cell activation once antigen receptors have been engaged.

Cell-mediated immunity/cellular immunity Adaptive immune responses initiated by antigen-specific T cells.

Chemokines A family of related small polypeptide cytokines involved in directed migration and activation of leukocytes. *Chemotaxis* is targeted movement in response to a chemical stimulus.

Class-switching A change in the production of antibody isotypes due to maturation of the B lymphocyte response to a particular antigen.

Cluster of differentiation (CD designation) A commonly used designation referring to specific cell surface molecules; useful in discriminating among cell phenotypes and assessing functional cellular activity.

Combinatorial joining The physical joining of nucleic acid sequences during development to generate novel proteins involved in antigen-binding receptors on B and T lymphocytes.

Complement A system of serum proteins involved in inflammation and immunity; mediates activities that include activation of phagocytes, direct cytolysis of target cells, and coating (opsonization) of microorganisms for uptake by cells expressing complement receptors.

Complosome An intracellular complement containing compartment-affecting normal cell physiology.

Concanavalin A (con A) Mitogenic lectin derived from the jack bean that stimulates T lymphocytes to undergo mitosis and proliferation.

CRISPR-cas9 A technology that employs enzymes to reliably make precise and targeted changes to the genome of living cells and organisms.

Cross-reactivity The binding of an antibody to an epitope or molecule that is similar in structure to the antigen used to elicit antibody response.

Cyclosporine A A powerful immunosuppressive agent.

Cytokine A class of small-molecule immune mediators secreted by leukocytes as a mechanism of immune regulation and cross-talk. Cytokines produced by lymphocytes are called *lymphokines* or *interleukins.*

Cytotoxic T cell A T lymphocyte that bears CD8 cell surface molecules that respond to antigenic stimulation through elicitation of toxic mediators; critical for antiviral and antitumor responses.

Danger-associated molecular pattern (DAMP) A conserved molecular motif associated with trauma and tissue damage that are able to trigger innate immune function.

Defensin A natural molecule able to limit the growth of microorganisms.

Degranulation A process by which myeloid leukocytes release digestive proteins stored in cytoplasmic vesicles.

Delayed-type hypersensitivity (DTH) *See* Hypersensitivity, Type IV.

Dendritic cell (DC) A primary phagocytic antigen-presenting cell (APC) that is capable of initiating immune response and lymphocytic activation, accomplished by cytokine secretion.

Enzyme-linked immunoadsorbent assay (ELISA) An assay used to detect antigens bound to solid wells in a plate format. Labeled reagents are used for quantitation by linking enzymes to an antibody to allow substrate color change to be used to recognize antigenic detection.

Endocytosis A mechanism utilizing receptors or pinocytosis whereby materials are taken up from solution into plasma-membrane vesicles by cells.

Eosinophil A polymorphonuclear granulocytic cell (PMN) involved in the innate response to parasitic infections.

Epitope An antigenic determinant; a portion of antigen capable of interacting with antibody or eliciting a lymphocytic response.

Erythrocyte Common blood cell (lacking nucleus) involved in oxygen delivery to tissue. Also called red blood cell or red blood corpuscle.

Extravasation A movement across blood endothelial barriers into tissue.

Fab fragment/F(ab)02 A portion of the antibody heavy and light chains that combine to make up the antigen-binding region.

Fas_FasL A cell surface molecule interaction that is required for the activation of apoptosis.

Fc fragment A portion of the antibody heavy chain that comprises regions that can interact with cellular receptors; confers biological function on the immunoglobulin.

Foreign Nonself.

GALT See MALT.

Germ line Genetic material in its original configuration, representing nonrearranged chromosomes.

Germinal center A secondary lymphoid tissue site where lymphocytic populations can proliferate and mature in response to antigen.

Graft-versus-host disease (GVHD) A state in which donor immune cells develop pathological reactions to recipient posttransplantation.

Granulocyte The general term for a phagocytic leukocyte containing granular particles.

Granzyme A protein involved in cytotoxic reactions and cell lysis.

Hapten A small, low-weight molecule that can elicit immune responsiveness only when attached to a larger carrier molecule, thus rendering it immunogenic.

Heavy chain A larger protein associated with the antibody molecule; confers biological functions and is associated with the constant portion of the chain.

Helper T cell A class of CD41 T lymphocytes that respond to antigens by secreting cytokine subsets to help cells to become effectors of cellular immunity or to stimulate B cells to make antibodies.

Hematopoietic stem cell A precursor cell found in bone marrow, which can give rise to leukocytes.

Herd immunity A social concept referring to preventing the spread of infection within a community; vaccination of a significant portion of a population provides a measure of protection for individuals who have not developed immunity, due to limitation of infection spread.

Heterograft A graft in which the donor and recipient are of different species. *See also* Xenograft.

Histamine A compound released from neutrophils during immunological and allergic reactions that cause vasodilation and smooth muscle contraction.

Histocompatibility Tissue compatibility between individuals based on the presence of polymorphic major histocompatibility molecules present on cellular surfaces.

Human leukocyte antigen (HLA) The genetic designation for human major histocompatibility complex (MHC) molecules. Class I molecules are represented by gene loci HLA-A, HLA-B, and HLA-C. Class II molecules are represented by gene loci HLA-DR, HLADP, and HLA-DQ. *See also* major histocompatibility complex (MHC) molecule.

Human immunodeficiency virus (HIV) A Lentiviral family member retrovirus with an RNA genome that forms a DNA intermediate incorporated into the host cell genome. Infection leads to the loss of CD41 lymphocytes and an eventual state of acquired immune deficiency.

Humoral immunity Refers to antibody-mediated immunity.

Hyperacute graft rejection A reaction representing immediate recipient antibody reactivity to antigens present in donor tissue during transplantation.

Hypersensitivity Immune reactivity to antigen at levels higher than normal, often leading to clinical states. Reactions are classified by mechanism: Type I (allergic) reactions involve immunoglobulin E (IgE) triggering mast cells; Type II (cytotoxic) reactions involve immunoglobulin G (IgG) against cell surfaces, resulting in cytolytic events; Type III (immune complex) reactions involve a destructive deposition of antibody and antigen complexes; and Type IV (delayed type hypersensitivity, or DTH) reactions are T cell-mediated.

Hypervariable region A portion of the antibody light and heavy chains that represents the most variable amino acid sequences coding for contact with the epitopes on the antigen; encoded by *complementarity-determining regions*.

Immune Checkpoint Point of regulation for self-tolerance to prevent attack of self.

Immune/immunity Protection against a specific disease or pathogen resulting from effective innate and adaptive resistance.

Immunization The induction of adaptive immunity by preexposure to antigens or by active infection, thereby generating a memory lymphocytic response.

Immunodeficiency A relative decrease in immune responsiveness due to the lack of components (innate or adaptive) capable of responding to a foreign influence. Immune deficiency disease is the resultant clinical state when parts of the immune system are missing or defective.

Immunogenic Capable of eliciting an immune response.

Immunoglobulin Any of five classes of antibodies [immunoglobulin M (IgM), immunoglobulin D (IgD), immunoglobulin G (IgG), immunoglobulin E (IgE), or immunoglobulin A (IgA)] that function in immune regulation through specific binding of antigens. *See also* Antibody.

Inflammasome An innate immune sensor that regulates caspases to induce proinflammatory cytokines.

Inflammation The build-up of fluid and cells that occur in responses to acute injury or trauma.

Innate immunity A component of the immune system consisting of genetically encoded constitutive factors that are readily able to respond to pathogens on short notice. The factors involved do not change or adapt during the lifetime of the organism; there is no associated memory response.

Interferon A specialized subset of cytokines originally discovered as having properties that interfere with viral replication; contains mediators of cellular immune function.

Interleukin *See* Cytokine.

Isograft Tissue graft between individuals of genetic identity. Also called "syngraft."

Isohemagglutinin A naturally occurring IgM molecule that recognizes ABO antigenic determinants on red blood cells (RBCs).

Isotype An antigenic marker that distinguishes members of an immunoglobulin class. Immunoglobulin isotypes include immunoglobulin M (IgM), immunoglobulin D (IgD), immunoglobulin G (IgG), immunoglobulin E (IgE), and immunoglobulin A (IgA).

Isotype-switching Genetic rearrangement in B lymphocytes to allow changes in the production of immunoglobulin isotypes.

Lactoferrin An innate iron-binding component with bactericidal and bacteriostatic activity, as well as immune modulating properties; found in neutrophils and secreted onto mucosal surfaces.

Leukocyte Any white blood cell (WBC; myeloid or lymphoid) that plays a functional role in either innate or adaptive responses. The myeloid population includes neutrophils, eosinophils, basophils, mast cells, monocytes and macrophages, and dendritic cells. The lymphoid group includes the lymphocyte populations.

Light chain A smaller protein associated with the antibody molecule; can be either the kappa (κ) or lambda (λ) variety.

Lipopolysaccharide (LPS) An endotoxin component of a Gram-negative bacteria cell wall that elicits mitogenic activity.

Lymphatic An endothelial-lined network of vessels permitting the flow of lymph to lymph nodes.

Lymphocyte A lymphoid-derived leukocyte expressing an antigen-specific receptor. There are two broad categories, T cells and B cells. Lymphocytes function as an integral part of the body's adaptive defenses and are critical to distinguish self from foreign antigens.

Lymph node A small, rounded, secondary lymphoid organ where mature leukocytes, especially lymphocytes, interact with antigen-presenting cells (APCs).

Lymphoid Tissue responsible for producing lymphocytes and antibodies, including regions in the lymph nodes, thymus, and spleen.

Lymphokine *See* Cytokine.

Macrophage A myeloid-derived cell involved in phagocytosis and the intracellular killing of microorganisms and antigen presentation to T lymphocytes.

Major histocompatibility complex (MHC) molecule A polymorphic molecule that allows the immune system to distinguish between self and foreign substances. Class I molecules present antigens to CD81 cytotoxic T lymphocytes and are on all nucleated cells. Class II molecules present to CD41 helper T lymphocytes and are found on antigen-presenting cells (APCs).

Mast cell A large myeloid cell found in connective tissues that mediates allergic reactions.

M cell A microfold cell found in the follicle-associated epithelium of Peyer's patches that functions to sample antigens from the small-intestine lumen to deliver via transcytosis to presenting cells and lymphocytes located on the basolateral side.

Membrane attack complex (MAC) The terminal product of complement cascade, whereby components C5 through C9 self-assemble on a membrane to form a cytolytic pore.

Mitogen An agent capable of stimulating cellular activation and division.

Molecular mimicry A cross-reactive occurrence during the development of autoimmune disorder in which a microorganism contains antigenic determinants that resemble those on self-tissues.

Monoclonal antibody An antibody derived from a single B cell that is specific for a single antigen.

Monocyte Part of the innate leukocyte population; a blood precursor to the tissue macrophage.

Mucosa-associated lymphoid tissue (MALT) A diffusion system of concentrated lymphoid tissue found in the gastrointestinal (GI) tract, thyroid, breast, lung, salivary glands, and skin. Related to gut-associated lymphoid tissue (GALT), which represents Peyer's patches found in the lining of the small intestines, and bronchus-associated lymphoid tissue (BALT), which represents aggregations of immune cells in the lower respiratory tract.

Myeloid Derived from granulocyte precursor stem cells in bone marrow.

Natural killer (NK) cell A small, granular innate cell derived from lymphoid progenitors, which is able to destroy tumor cell targets rapidly by antibody-dependent cell cytotoxicity, which permits target destruction in a nonphagocytic manner. This type of cell does not express a T-cell receptor (TCR).

Natural killer T (NKT) cell A small subpopulation of T cells that express a limited T-cell receptor (TCR) repertoire; receptors recognize bacterial lipids or glycolipids bound to nonclassical histocompatibility class I-like molecules.

Neutrophil A polymorphonuclear (PMN), phagocytic granulocytic cell involved in the acute inflammatory response to pathogens.

Neutrophil extracellular trap (NET) A mechanism used by neutrophils to release a meshwork of DNA and chromatin fibers, forming extracellular structures to immobilize and kill microorganisms.

Nitric oxide A molecule important in intracellular signaling; a free radical and regulator of hydrogen peroxide in phagosomes within phagocytic cells.

Opsonization The process by which a molecule or pathogen is targeted for ingestion and subsequent destruction by phagocytic cells, mediated through complement or antibody interactions. An *opsonin* is a molecule that enhances directed phagocytosis.

Paratope The portion of the antibody that contacts the epitope on the antigen.

Passive immunity The state of immunity acquired through the transfer of factors (serum or antibodies), which allows a protective state in the absence of active immunity.

Pathogen-associated molecular pattern (PAMP) A conserved molecular motif associated with infectious agents that are able to trigger innate immune function. Cellular receptors on monocytes that recognized conserved molecular motifs associated with infectious agents are called *pattern-recognition receptors (PPRs)*.

Pentraxin An innate pattern recognition molecule that functions to bind debris during infection and inflammation.

Perforin A protein involved in cytotoxic reactions and cell lysis.

Peyer's patch A lymphatic nodule located along the small intestine.

Phagocyte Any mobile leukocyte that engulfs foreign material. The process of directed uptake is called *phagocytosis*.

Phagolysosome An internal digestive compartment within phagocytic cells where phagosome and lysosomal enzymes destroy engulfed pathogenic invaders and digest engulfed proteins.

Plasma The fluid component of blood, containing water, electrolytes, proteins, and molecular mediators; plasma does not contain cells.

Plasma cell A terminally differentiated, antibody-secreting B lymphocyte.

Platelet Blood cell (lacking nucleus) involved in hemostasis, preventing bleeding. Also called thrombocyte.

Polymorphonuclear cells (PMNs) A group of white blood cells (WBCs; neutrophils, basophils, and eosinophils) with multilobed nuclei and cytoplasmic granules.

Primary immune response An adaptive immune response representing initial exposure to antigens, predominantly comprised of immunoglobulin M (IgM), followed by the later presence of other antibody isotypes. *Priming* is the activation of lymphocytic response to antigens for the first time, initiated by antigen-presenting cells (APCs).

Primary lymphoid tissue Immune organs where lymphocytes develop and mature; organs where antigen-specific receptors are first expressed.

Regulatory T cell (Treg cell) A specialized T lymphocyte subgroup that is able to regulate immune responses and is effective postthymic development.

Respiratory burst Phagocytic metabolic activity resulting in the formation of superoxide anion and hydrogen peroxide.

Rhesus antigen (Rh) *See* Blood group antigen.

Rheumatoid factor An immunoglobulin M (IgM) isotype antibody reactive with immunoglobulin G (IgG) molecules.

Secondary immune response An immune response induced by repeated antigen exposure, often of higher affinity and with greater speed than elicited by primary response; has the characteristic maturation of antibody isotypes.

Secondary lymphoid tissue Immune organs where antigen-driven proliferation of lymphocytes occur in response to antigenic stimulation.

Secretory component The portion of the dimeric immunoglobulin A (IgA) molecule that is critically involved in release across mucosal barriers.

Seroconversion Indicates when antibodies can be first detected against antigens following infectious challenge or immunization.

Severe combined immune deficiency (SCID) A disease state in which defects in maturation pathways for both B and T lymphocytes result in the lack of functional adaptive immunity.

Somatic hypermutation A change in affinity maturation of the antigen-binding site in an antibody following antigenic stimulation.

Superantigen A molecule able to elicit T-lymphocyte responses by circumventing normal antigen-processing and -presentation functions.

Syngeneic Being from individuals that are genetically identical.

T cell/T lymphocyte A cell derived from a bone-marrow, lymphoid progenitor stem cell, possessing specific cell-surface antigen receptors; types include helper T cells of different cytokine secreting subsets, as well as cells that confer regulatory and cytotoxic function.

Titer Relative antibody concentration.

Tolerance A state of less responsiveness to a substance or a physiological insult; instrumental in the prevention of autoimmunity.

Toll-like receptors (TLRs) A subset of pattern-recognition receptors (PRRs) recognizing conserved molecular motifs associated with infectious agents; initiate strong innate immunity when triggered.

Thymocyte A hematopoietic progenitor cell present in the thymus.

Vaccine An immunogenic substance used to stimulate the production of protective immunity (antibody- or T cell-based) to provide protection against clinical disease.

Vaccination The artificial induction of adaptive immunity by the preexposure of antigens or pathogens to generate a memory lymphocytic response.

Variable domain/variable region The end portion of the antibody or T-cell receptor (TCR), which comprises the antigen-binding region.

V(D)J Recombination Mechanism of genetic recombination in developing T and B lymphocytes during maturation. Results in production of B cell receptor (antibody) or T cell receptor (TCR) for antigen recognition.

Western blot A diagnostic antigen identification of a mixture separated by electrophoresis through a gel matrix. Proteins are transferred to a solid matrix (usually nitrocellulose) and probed with specific immune reagents.

Xenograft A tissue graft in which donor and recipient are of different species.

Index

Note: Page numbers followed by *f* indicate figures, and *t* indicate tables.